Trust in Virtual Teams

This book is dedicated to my dear wife Nance, and my children. It is their support, dedication and encouragement through the years of my education, experience, and research, and through hours of isolation, late nights, and early mornings that makes this work possible. Everyone who has attempted their education as a father, husband, mother, and wife knows what I mean. I want to thank you all, and hope that each of you may be encouraged as you work toward your dreams.

Trust in Virtual Teams

THOMAS P. WISE

Routledge
Taylor & Francis Group

LONDON AND NEW YORK

First published in paperback 2024

First published 2013 by Gower Publishing

Published 2016 by Routledge
4 Park Square, Milton Park, Abingdon, Oxon OX14 4RN

and by Routledge
605 Third Avenue, New York, NY 10158

Routledge is an imprint of the Taylor & Francis Group, an informa business

Publisher's Note
The publisher has gone to great lengths to ensure the quality of this reprint but points out that some imperfections in the original copies may be apparent.

British Library Cataloguing in Publication Data
Wise, Thomas P.
    Trust in virtual teams : organization, strategies and assurance for successful projects.
    1. Virtual work teams. 2. Project management. 3. Psychology, Industrial. 4. Trust. 5. Quality assurance.
    I. Title
    658.4'022–dc23

The Library of Congress has cataloged the printed edition as follows:
Wise, Thomas P.
    Trust in virtual teams : organization, strategies and assurance for successful projects / By Thomas P. Wise.
        pages cm
    Includes bibliographical references and index.
    ISBN 978-1-4094-5361-1 (hardback : alk. paper)—ISBN 978-1-4094-5362-8 (ebook)—ISBN 978-1-4094-5362-8 (ePUB)
    1. Virtual work teams. 2. Virtual work teams—Management. 3. Teams in the workplace. 4. Trust. I. Title.

    HD66.W57 2013
    658.4'022—dc23
                                                        2012045314

ISBN: 978-1-4094-5361-1 (hbk)
ISBN: 978-1-03-283744-4 (pbk)
ISBN: 978-1-315-54962-0 (ebk)

DOI: 10.4324/9781315549620

Contents

List of Figures

List of Tables

About the Author

Dr Wise is a Ph.D. in Organizational Management from Capella University with specialization in Information Technology Management currently teaching part time at Villanova University in the MS Engineering Computer Science program, and part time with DeSales University in the MBA and MSIS program. As Director of Quality Management for a major communications company, Dr Wise is responsible for process improvement, problem solving, internal consulting and assessments, as well as quality assurance and metrics and reporting at the National Engineering and Technical Operations National Test Lab in Pennsylvania, USA.

Dr Wise is a member in good standing with the American Society for Quality (ASQ) in Millwaukee Wisconsin, USA, and a ASQ certified Manager of Quality and Organizational Excellence since 1997. Wise is also a certified Management Professional since 1998 with the Project Management Institute Project located in Newtown Square Pennsylvania, USA. His education includes a BS Organizational Management and a dual major in Human Resource Management from the College of St. Francis, Joliet, Illinois, USA, and a Masters in Business Administration specialized in management from the University of St. Francis, Joliet, Illinois. The author's work experience includes work as an internal quality consultant with a major commercial nuclear power electric production company, and the development of quality programs for major financial market producers. Dr Wise currently resides in the Philadelphia, Pennsylvania area with his family.

As a quality professional in the industries of commercial nuclear, financial, and mass media, experiences in project and quality management have led to an interest in project, software and systems development, and quality, and the effect these processes have on the human interests of trust and communication. Of particular interest, and current research, is understanding the effect of a virtual work environment in the areas of project, systems and software development, and quality management has upon trust and communication. Current research includes trust in virtual project teams in relation to the variable employee role and the elements of virtuality.

Foreword

Teamwork on a guiding change coalition can be created in many different ways. But regardless of the process used, one component is necessary: trust. When trust is present, you will usually be able to create teamwork. When it is missing, you won't.

John P. Kotter

As an organizational transformation specialist, managing organizational change is an everyday affair for me. It didn't take long to realize the truth behind Kotter's edict. Accomplishing that realization is usually the first challenge in any transformation initiative. In these days of high speed communications, globalization and the relentless drive for productivity, most companies have implemented virtual teams in one form or another. Developing trust for teams in such an environment is no longer an exception. *Trust in Virtual Teams* addresses this necessity. It also fills a huge void in our business libraries on this topic. Dr Thomas Wise uses his rich experience in handling teams in this book. His diverse experience includes managing trade union groups, offshore teams, and modern day virtual teams. This versatility is revealed through the scenarios and examples featured throughout the book.

According to Dr Wise, information (or sometimes the lack of it) affects our perception of trust and our judgment of corporate communication. After describing how personality, cognition and institution affect our perception of trust, Dr Wise, provides some innovative ideas to significantly improve institutional trust within teams. He establishes that understanding QA process is critical to build a clear understanding that the policies and procedures of the organization are equally and fairly applied to all team members.

In my experience, I am seeing a growing trend for organizations to be agile and lean. As these organizations transform to a more efficient future state, the importance of trust and communication is all the more significant. A firm with a fragmented team will never be able to survive in this day of competition. Dr Wise, in his scholarly appeal, underlines the need to implement practices that enhances institutional trust through compliance. This best practices handbook is long overdue in our business community.

The book is well structured and illustrated with examples and figures that makes this great reading for any business leader. It is undoubtedly the best guide available for understanding trust in virtual teams. Business practitioners will enjoy the book because of the many case studies and examples cited. *Trust in Virtual Teams* is a valuable read for any manager who is involved in managing virtual teams.

Reuben Daniel
Global Transformation Leader, Cognizant

Preface

Have you ever tried to write something original? When sitting down to write, and staring at a blank page, the mind slows to a snail's pace, the pulse quickens, and the information flow halts. Nothing happens, and the writer begins to question that there is anything new to be said. The very same result occurs when a manager works to begin a program designed to increase trust in order to increase operational readiness. As I began to write this book, I spent weeks staring at a blank screen, beginning to write, then erasing, starting again, and erasing.

So I thought, maybe I should start in the most logical place to describe trust.

In the beginning.

When we were little we got a lot of advice. People were always telling us to be careful, look both ways before crossing the street, and wear clean underwear. Look before you leap. Don't J-walk, and make sure your socks match. The most important of all advice that anyone ever received was NEVER trust *strangers*. Advice was easy to get, easy to give, and forever present when we were kids, but now, when we need it most, advice has changed.

Advice now comes in the way of high priced consultants willing to write long, detailed, and complicated power points that leave us hanging, and wanting more. It is complicated, hard to implement, and designed mainly to sell us more advice. What we need is more of what Mom told us. We need simple, clean, and straightforward advice about life that we can reach back and pull to the front.

Trust is simple, and yet one mistake can set us back on our heels for a very long time. Trustworthiness can dissipate in a moment of hesitation, a slip of the tongue, or errant email that escapes with a wrong key strike. A perceived lack of trust can be precipitated by the vast gaping maw of a silent cloud called the Internet as we wait anxiously for an important, and never arriving reply, or a missing report. The always present appearance of favoritism can drive a deep wedge between people and teams.

If only someone would simply remind us to be careful, look both ways, and wear clean underwear life would be once again good. I always feared the stranger the most. For some odd reason I could never be sure who exactly the stranger was.

Why I Wrote This book, and Perhaps a Little Guidance in How to Proceed

As I have worked to explain to several international corporations in the financial and communications industry, the materials available do not begin to provide a clear understanding of how a simple concept like quality assurance greatly impacts institutional relationships in a virtual environment. Virtuality is not understood in business. Business, even those with completely collocated teams, work virtually.

Virtuality is found in how team members work, and not in where team members sit. Because today's team members work in virtuality as much by choice as by geographic separation, business leaders must understand how relationships such as project team trust, cross-divisional projects, and offshore team participation are all positively motivated by a solid quality assurance program. This book provides a clear view of what virtual projects can be, and how quality assurance makes or breaks trust and relationships in a virtual environment.

Feel free to start reading in the very first chapter to begin building a quality assurance program that supports trust, both within your project teams, and within your individual employee relations. If, however, a better understanding behind why quality assurance may function to provide a strong trust environment, begin reading in Part III. This section of the book provides some background in the lesser known aspects of quality, and some in-depth information regarding recent research and the effect of corporate policy in support of project team trust.

For those who have an interest in the most recent research regarding project roles and the effect of virtuality on those roles, Part II is a great place to begin reading. After reading this session, the reader will have a good foundation in the most up-to-date knowledge on trust within virtual project teams. Managers may use this information in building effective strategies to support trust and build highly effective virtual work environments.

Acknowledgements

Researchers and educators have a saying that reads something like this. Each of us stands on the shoulders of those who have gone before us as we seek to understand, and add to the body of knowledge. This work is no different. All of those who are cited here in, and those whose research guided my experiences and education, should be acknowledged for their contribution. Those with whom I work, now and in my early years in the power plants and in the financial exchanges, are given my thanks. Your encouragement, and challenges, over the years are greatly appreciated. I think of you often with fond memories.

This book represents many years of learning and research, both secondary and primary, and the experiences as an internal consultant and line manager in many of the practices in organizational, IT, and product quality. We all struggle with understanding the full capability of a robust and healthy quality program. I hope this book helps to reignite an excitement in the quality practitioners and consumers alike regarding the possibilities that a fully engaged quality program brings to the organization.

I would like to especially note, and thank, Janai Wise for her work, and her guidance on the illustrations and diagrams used in this book, and hope that her contribution encourages her work in professional art. I would like to acknowledge the support, love, and encouragement of my wife Nance, and my children for their smiles, hugs, and enthusiasm for this project. You have no idea how important hugs and smiles can be. Finally, I would like to acknowledge the words and prayers of my parents over the years as I worked down the many goals that carried me to completing this work. Lastly, and yet never least, I would like to acknowledge the patient guidance of my editor, Jonathan Norman.

List of Abbreviations

API   Application programming interface

CumCnt   Cumulative Count

CumPct   Cumulative Percent

IM   Instant messaging

INPO   Institute of Nuclear Power Operators, USA

IS   Information Systems

IT   Information technology

NRC   Nuclear Regulatory Agency, USA

OSHA   Occupational Safety and Health Administration, USA

PMO   Project management office

QA   Quality assurance

QC   Quality control

RAD   Rapid Application Development

SMART   Specific-Measurable-Attainable-Relevant-Time bound goals

SME   Subject matter expert

TQM   Total quality management

VPN   Virtual Private Networks

PART I

Understanding and Building Trust

Introduction

Why are communication and trust once again topping the list of your annual employee survey? Every year executives struggle to determine why their employees don't trust the company, and why the employee survey consistently says communication is the biggest problem facing employees. Teams are formed each spring to address these very problems, and at the end of the year trust and communication once again top the list. Trust and communication are often one and the same, in the sense that they are mutually interdependent.

An individual's awareness of trust grows in layers. Beginning in our early years, as our personality starts to form, we develop a sense of what trust means through our experiences. When expectations are met we learn to expect that those same expectations may be met in the future, and as we watch the world around us we build within us a sense of equity, of rightness or fairness. As we experience life as an employee we learn fairness and equity through our perception of company policy and the implementation of those policies, and as we develop as team members and group participants we discover information about our work life and peers.

Each of these layers of experience create our expectations, our values, and the basis on which we trust one another. We carry to the workplace our trustworthiness and an expectation of the trustworthiness of others, as well as an intuitive ability to measure and prescribe trust upon each other, and assess the balance of equity. In a 2003 study Sarker, Valacich, and Sarker described these layers of trust as personality based trust, institutional based trust, and cognitive based trust. Information (or sometimes the lack of it) affects our perception of trust and our judgment of corporate communications.

When we understand trust in the collective form, we can then understand trust as situational to the given conditions. Trust, when we are talking about personality based trust may be dependent upon the individual's experiences from early childhood regarding the trustworthiness of specific stereotypical descriptions based on role, race, or ethnicity, or given conditions. This can be in part based on one's personality traits such as suspiciousness regarding a belief in the inherent honesty of society. It is the discovery of information, and the development of effective communication strategies, that makes trust and communication come together to form a solid bond of connectedness that underpins an effective team and company culture (Halgin, 2009). A sense of family, or connectedness, provides for the expectation of a shared future in their membership, lowers the sense of vulnerability in the group, and opens channels for self-disclosure.

When we look at cognitive based trust, we are evaluating information that is discovered regarding the situation or persons involved in a specific situation, or perhaps forming conclusions regarding the trustworthiness of a group. A decision to trust or not trust is consciously evolved based on an analysis of the available information. As team members seek information, the degree to which reliable information can be discovered will hamper or enhance the degree to which team members may conclude that trust is worthy in the situation.

Institutional based trust describes the level of trust that team members may have in the degree to which corporate policies and procedures are fairly and honestly applied across the organization. As team members work in and with organizations and individuals, they will draw conclusions regarding the way in which management implements and manages the processes and policies governing their work and work lives. Perhaps the establishment of a well supported quality assurance function may provide the needed transparency in process reporting regarding adherence practices for institutional rules. The primary focus of Quality Assurance (QA) as an organizational process is the discovery and analysis of information, and the effective packaging and reporting of this information to make it easily discoverable.

When QA is done properly, the focus of the QA program is aligned to the strategic plan, and the tactical goals of specific organizations. The process of QA enables effective discovery of information, makes transparent our expectations on compliance to company policy, and focuses the organization on what is important. Tracking and reporting on how well we are doing against our own policies builds solid institutional level trust, establishes the foundation for cognitive based trust, and capitalizes on our own personality based traits regarding expectations of trustworthiness. It does not do us any good to track and manage our business processes if we fail to share the results of this analysis. We want everyone to know how well we are doing with compliance. Ironically, poor compliance can build trust, just as effectively as good, but it is good compliance that builds strong bonds.

If we are collectively able to trust the information we receive as accurate and fair, then simple reports, auto-generated to tell everyone how well a project is complying with the stated lifecycle, are all that an effective organization requires. These reports can be held in a standard repository that everyone can access for all the projects and can use to get immediate feedback on the health of the business. This type of openness and ease of discovery can provide effective and immediate feedback to all employees as to the equitable implementation of company policies. When done well, it provides quantifiable evidence that we are equally entitled to and share a common set of practices no matter where we sit. Trust is supported by common norms and practices, and while very basic, the practices that define how work gets done are very important.

A key to effective reporting is to ensure the data collected and analyzed are designed to answer the questions that people are asking. All too often QA programs are maligned and eventually abandoned because the program collects data that is easy to collect rather than data that is necessary, and present the data in expensive, hard to maintain charts. Adding to the problem is the often used practice of presenting the data in a location to which many people cannot find, and making the charts hard to identify and locate. Before beginning any QA program, first ask yourself what questions need to be answered. What is the dilemma to be resolved, and what questions must management answer to resolve the dilemma. Next, identify specific research questions that will provide the data to answer the management questions. If the QA program begins with these simple points in mind, the data will be both useful and effective in providing information that is well designed, and will likely support the needs of the organization and the individual team member.

The goal here is to build a clear understanding that the policies and procedures of the organization are equally and fairly applied to all participants. By making public how the lifecycle is managed, and the degree to which participants in the lifecycle are expected to comply with the policies and procedures, all participants are able to gather the information to confirm an even playing field. But what happens if our publicly displayed information shows that the process is broken or that no one is following it?

Understanding Trust

Trust grows in layers, beginning in our early years. As our personality develops we develop a sense of what trust means through our experiences with trustworthiness. When expectations are met we learn to expect that those same expectations may be met in the future, and as we watched the world around us we built within us a sense of how it all works; of equity, or fairness. Each of these layers provides the basis on which we trust one another. This holds true for the trust between employees, teams, departments, and organizations. We carry to the work place our trustworthiness and an expectation of the trustworthiness of others, as well as an intuitive ability to measure and prescribe trust upon each other, and assess the balance of equity. In a 2003 study, Sarker et al. described the layers of trust as personality based trust, institutional based trust, and cognitive based trust.

Personality Based Trust

Our personality, I am told, is pretty much set by the time we reach the age of seven. It means that when it comes to personality based trust the company receives the employee's personality as fully matured, and with limited opportunity for shaping. Personality develops as the result of experiences during the earliest times in our lives. The other kids were always more than willing to tell us who we should trust: don't trust Jimmy (he's a liar), and always trust Sara (she's a friend). Trust in these circles was based on who knows who, and who follows through on their promises, or meets our expectation of trustworthiness. Thus the expectation of trust is pretty much established by the third grade.

As we grow, and our experiences branch out to include school and work, consistent behaviors begin to breed new levels of trust. We can trust in behaviors that are consistent, even when those behaviors are consistently threatening, because the consistency provides a backdrop for trust to develop.

As we grew we all knew kids that the other kids and parents would tell us we should stay away from, and which kids we could approach. Kids know they can trust that a bully will always behave like a bully, even when at times they appeared to want to be friends. The bully could always be trusted to be the bully in the end. Thus, for most of us, inconsistent behavior amongst friends is far more confusing and, indeed, threatening, than the consistently threatening behavior of the bully. Our expectations and desire for consistency and strong relationships become key to personality based trust at the corporate level.

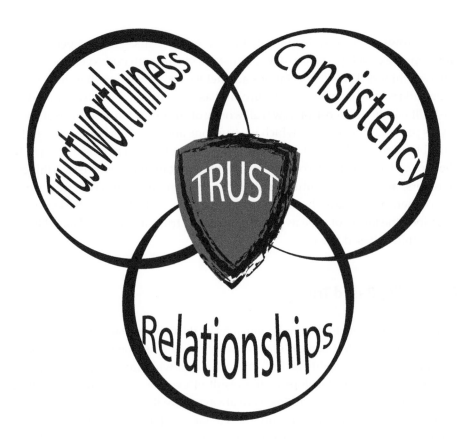

Figure 1.1   Framework of personality based trust

These experiences will often define our expectation of the work place and affect how we see our co-workers, employees, bosses, and team mates. In each of us there lies a propensity to trust, or not, which affects how we perceive the trustworthiness of others. In us we carry the expectations of what is trustworthy, as we also develop a sense of our own trustworthiness. Personality based trust, as Figure 1.1 depicts, can be said to be a juncture of perceived trustworthiness, the consistency with which the person acts as expected, and the relationship with the trusted person. When the three points of trust come together, personality based trust may be prescribed upon the trusted person. At times this may seem odd, but the bully may be trusted; at the very least, trusted to be a bully.

Cognitive Based Trust

We learned to watch out for the bully, and we learned who our friends really are in a pinch. Learning creates an understanding of the world around us. Cognitive trust is a trust that we choose to place in a person, group, or program based on information that we have gleaned from our past. This may include information learned during our years of education and learning; perhaps through experiences we have encountered in similar situations with people or programs. As we take our past learning for use in current or similar situations to decide if trust is a reasonable response to the situations we face, the learning is often enhanced if we have a solid belief that we may also have a future relationship with the individual or group (Mizrachi, Drori & Anspach, 2007). The expectation of future relationships with a good experience have a way of strengthening that relationship, because when we expect that a trusted condition or relationship will remain, it will cause reliance upon that condition or relationship to grow.

As we grow, our experiences in the past become our expectations for the future in terms of situations or behaviors of the people around us, and we don't need to look further for evidence to decide to trust. Remember the saying, "familiarity breeds contempt?" This may have some truth to it. Familiarity, according to Lewis and Weigert (1985) is a precondition to prescribing or withholding trust. I often remember how, when I was told not to trust another kid, I would wait and watch, wondering what they might do. As I learned more about that person, then I would learn to trust that he or she would act in a certain way, in a given type of situation. As a kid I played a lot of sports. Growing up in a military family in the US the expectation was that all the guys played organized baseball, football, and basketball, depending on the time of the year. Often one of the dads coached the team. The first day of practice we would all gather at the field watching as each of the kids showed up. One at a time each of us would emerge from their car with a parent, and the rest of us would watch, waiting to see who would approach carrying a rucksack filled with sports equipment. As the designated dad-coach showed himself we would keep a close eye watching for some clue as to how to build our expectations for the season.

As the dad-coach would emerge with the rucksack full of equipment the discussion would immediately begin. Questions would fly around the team gathered at the field. Has anyone ever been on his team in the past, someone would immediately ask? We would either hear excited responses from those with good past experiences, or groans and warnings from those with bad past experiences. From the stories, the first set of inputs in everyone's decision to trust or not to trust was the past experiences from anyone with something to share. Team members would share stories from friends and past team mates about the coach's reputation for fairness, and coaching style. We would all keep an eye on the coach to see who had joined the entourage as they all emerged from the car. Once again groans and moans or excitement and anticipation would pass among the crowd based on the company that the coach kept. The reputation of those accompanying the coach would have an immediate effect, either positive or negative, on our decisions to trust in our coach. If the coach's kid was a solid athlete and had a reputation for hard work and fair play, that too would color the thoughts of everyone gathering information regarding the new coach. The next input was the first words from the coach's mouth, and we all waited and watched as the coach approached the group. Often a coach would shout as they arrived for everyone to stand and begin running laps, letting us know immediately that who we were was not of the first importance. The coach that arrived with a vision for the season, a roster in hand showing that he had an interest in who we were and maybe some idea of what we had done in the past, and perhaps a schedule for the day and the rest of the season, would set a tone of consistency and planning. Each of these inputs provided answers to our information search, and data to our questions regarding the trust relationship.

All of this set in the minds of our team mates whether or not we should keep a wary eye, or trust the coach to take good care of us. School would bring the very same watch and wait attitude from each of us. We gathered in a class room watching for the teacher to arrive as we talked amongst ourselves about who the teacher was, who already knew them, and what our mutual expectations might be regarding timeliness of assignments, rewards, and consequences. In both sports and school we learned from experience whether we could or should trust, and then decided individually if we would choose to trust that the coach or teacher would perform the same in the future. The prescription of trust in the cognitive model works the same in our teams and groups in the work place.

Trust is a choice based on data gathered from many sources. Our own experiences with the past behaviors we encounter, the current behavior, and whether or not we believe there may be a future in the relationship. As shown in Figure 1.2, a combination of experience, knowledge, and expectation of the future relationship are combined to aid the choice as to prescribe trust upon the person.

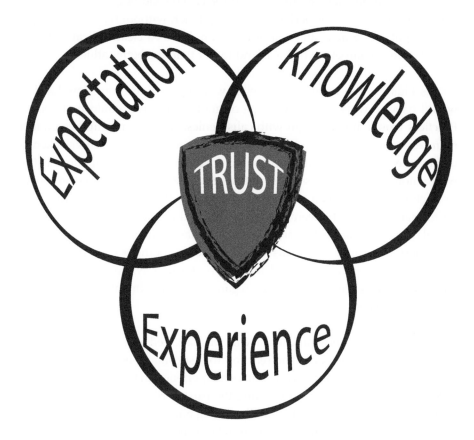

Figure 1.2   Framework of cognitive based trust

Institutional Based Trust

Institutional based trust is somewhat different. As we grow we come into contact with the administration of many different institutions. To keep with the youth sports analogy as my first experience in institutional trust, one of the first things any group seeks to identify is whether everyone is truly equal in the treatment and application of policies and practices. As the team would gather we would watch for clues as to whether the coach's kid would be treated the same as every other kid on the team. Would the coach's kid have to work as hard as the rest of us in order to get a starting spot in the game? The team would get an early indicator as practice began, whether or not the coach was going to play favorites based on whether his kid was automatically assigned one of the highly coveted positions on the team. If the coach's kid began practices as the starting running back, or perhaps the lead pitcher, without having to compete for the spot the same as the rest of the team, it was a clear indicator that we would have to compete to be his friend rather than the better athlete.

As kids do, the team would begin working to figure out how hard we would have to work, and what type of goofing around would be tolerated. Practices would always begin with calisthenics. As the warm up began, team members would immediately be watching to see if the kids who slacked off during the exercise would be chastised for their weak participation. If they were ignored, then of course those with a lesser level of motivation would begin to slack off to bring their performance down to the least level of acceptable performance. All of the team members watched as the coach interacted with the players, assigned roles, and doled out chastisement and punishment. If weak performance or effort was not addressed or perhaps even rewarded, or perhaps rules were eased to aid favorite performers while at the same time strong effort was ignored, attitudes regarding fair treatment from the leaders would be adjusted.

Over time everyone would learn what was expected of us, and what we could expect from the dad-coach. The reaction to institutional equity, or fairness, is the same in the work place as it was in our experience on the childhood practice field. We make our decision regarding the trustworthiness of an organization based on how consistently and equally among our peers we feel we are treated. As employees and team members we build our sense of institutional trust based on the treatment we receive in regard to the organizational policies and on how equitable our treatment is when compared to the treatment that others receive.

Later in life, I began to wonder more about how I would apply the rules when I was in charge. Should everyone to be treated the same? It seemed that, while I watched my boss, people that worked for me must also be watching me. My subordinates would be trying to decide if the rules truly are the same rules for everyone, and if everyone truly is to be treated equally.

My career grew until I was managing a group of very talented technical procedure writers in a nuclear power plant. Our job was to apply the Code of Federal Regulations, Nuclear Regulatory Guidelines, Occupational Safety and Health Administration (OSHA) Rules, and Original Manufacturer Recommended Practices equitably. Our work was regularly audited by the Nuclear Regulatory Agency (NRC) and the Institute of Nuclear Power Operators (INPO). Both the NRC and INPO would audit our practices to see if the rules established by the government agencies and manufacturers were properly and consistently applied. Again, when the rules were applied with consistent rigor, then the regulators gained trust in the company.

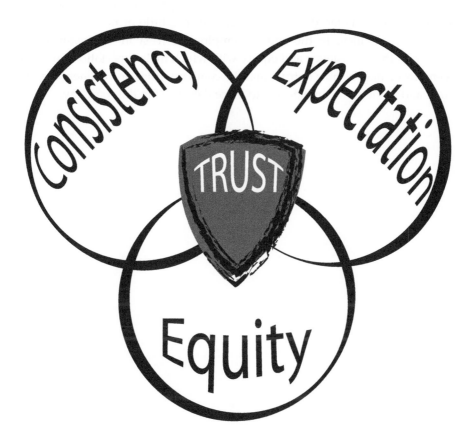

Figure 1.3   Framework of institutional based trust

If the company applied the rules equally to everyone, and in all situations, then my sensibilities told me the rules and the company could be trusted, even in cases when the rules seemed unfair. Fairness raises an interesting point regarding how fairness is perceived by employees. Fair is often defined as the absence of bias, and in the case of institutional bias, this is often interpreted by an employee as a balance between the value received and the value contributed. Recent research suggests that an employee's perception of fairness may also be influenced by the employee's personal perceptions of job satisfaction, and as a response to positive discretionary activities by the company. This may be due to the potential for positive discretionary rewards to have a symbolic relationship with trust, potentially balancing the problems of perceived bias due to the problems faced when management has little or no influence over organizational policy. In the end, an employee's trust in the institution is based upon the consistency and equity with which their expectations are met. An employee may be able to trust that the company rules or treatment may be unfair, and at the same time decide to place their trust in the company.

2

Why Talk about Trust?

Trust is essential in team formation and considered by some to be the basis of the Tuckman team development model formation phase, and an essential element in the development of effective virtual teams. When managing knowledge workers, Glen (2003) tells us, it is often hard to tell when they are working and when they are taking a break. Work in the modern office may include hours of confined, heads down, contemplation, followed by bursts of energetic code development, project outlines, phone calls, meetings, and online chats. In classic management style, we as managers are asked to make a determination as to whether value is added by team members whose value is in the form of abstract knowledge content. Team members participating as knowledge workers will often form a large portion, and in some cases the entirety, of many project teams. Further, with workers capable of contributing from anywhere in the world, participating in a virtual work environment, trust is often the only link between team mates.

A project team will have many cultures, many roles, and many personalities. As teams form they will seek ways in which they can gather information about their peers. This information will likely include their first impressions and the impressions of others as they try to develop a sense of similarities and differences. Team members will begin to form a set of desires regarding sub-groupings around attributes such skills, patterns of behavior, as well as tangibles such as task preferences and responsibilities. In order for this to happen effectively team members will need a degree of trust to support the information discovery phase that begins in team formation. The team formation phase is often complicated by project variables such as complexity, scale, and geography.

While in the early stages of team development, team members will tend to avoid the difficult questions and conflicts. As teams begin to take shape, members learn to interact with one another and team processes and roles mature. During this maturing process team members will find that conflicts will come. Priorities and project complexity will change with the level of team maturity and growth increasing the potential for conflict. Individual and team contributions will vary in complexity and size, as well as priority, and with these variables often come conflict and questions. Managers must be able not only to manage the questions and fears that arise, but also to avoid the unnecessary interruptions to project delivery that may come with them. In order to move the team beyond this stage, relationships must form to effectively deal with and resolve conflicts. Establishing both a trust relationship and an effective virtual work environment, based on that trust, is critical to reducing project risks.

In the modern office, although *virtuality* has traditionally been defined by distance, it is not necessarily dependent upon distance. All of us have experienced co-workers sitting on the other side of a cubicle wall working silently, and separate, and yet together on a project deliverable. The silence can be deafening as the peers and team mates use instant messaging (IM); build expert MSPowerPoints and MSVisio diagrams and email their work to one another to share their thoughts. In many cases the output of the project may be the gathering, analysis, synthesizing, and rendering of complex knowledge in a digestible form necessary to support or even define the elements of another dependent, and yet just as complex project.

Responses and thoughts relay from one knowledge contributor to another, building in complexity as the project priority shifts, and therefore leadership roles and participation levels change. Project outputs in the form of electronic documents such as MSVisio are instantly shipped to the next contributor to land upon their virtual desk top somewhere in the world. This somewhere may be one floor above or below, a cubicle row to the left or right, or perhaps across two time zones as the project progresses. With the capability of follow the sun project contributions, a single project thread may follow a 24-hour cycle to continue processing in London or Philadelphia where the process is once again picked up and repeated, passed across the office floor in silence, then on to Chicago and Los Angeles and Hong Kong as the sun moves westward.

Even very simple projects can be quickly derailed if there is a breakdown of trust in this virtual world. A project, by a commonly held definition, is an undertaking by one or more individuals that shares the basic elements of contemplation and planning, requiring resources of investment, people, and materials. This undertaking must be accomplished within defined constraints that include a specific beginning and end date. Project constraints are defined by project stakeholders, which may include investors, corporate leaders, product owners, and end users, and project managers and team members of dependent projects. Most projects have, based on the current paradigm, a named leader or project manager who must coordinate the basic elements of the project and provide guidance and delegation in order to accomplish the defined tasks.

When a manager seeks to delegate, the very definition of delegation requires that the manager release to the delegate the authority to accomplish the task within the defined constraints. The act of delegating is the act of releasing or sending a representative in your stead to act upon the desires of the sender. This act of sending one's representative requires a high degree of trust in the capabilities and integrity of the one sent. The project manager in this case is the sender, therefore becoming an enabler, identifying and clearing the way for the team's success.

Often, when project managers first begin to work with new teams, communication with this team may tend to be short, direct, and demanding, and may not leave any room for questioning and clarification. Sometimes, and perhaps due to the constraints in time and resources, project managers are playing a dual role in the project. In a recent study it was found that approximately 88 percent of project managers split their time between their project manager responsibilities and other roles such as developer or analyst. With demanding time restraints that may result in splitting time and dual roles, trust becomes an even greater element in the project environment.

Information may become unidirectional, and silence may be believed to imply understanding. As trust develops team members are more willing to ask the hard questions of one another. They become more willing to challenge the basic paradigms and find new and potentially better ways of getting work done, and become creative in problem solving. As senders and delegators build a strong trust relationship with their team members, and team members believe they have the authority to act and make decisions, they are no longer constrained to doing exactly what the project manager would ask, but rather are enabled to do what is needed. They become capable and willing to go beyond explicit direction.

Managers are often building teams with actors they may know only by name and reputation, or possibly, simply by position. Team members join a project as need for specialized skills arise, and often leave the team as their work completes. Team members may join a project from many different locations around the world. Each of them bring to the table differences in family culture, education, training, and past experiences, as well as differing levels of language skills and familiarity with organizational procedures. What may become a continuous stream of new names and personalities, while necessary to maintain progress and relevant skills, can become an additional challenge in building an effective team. Each time a change occurs in the team structure or membership, the potential for the team to regress in maturity does exist.

Trust, therefore, becomes an essential element in team resiliency. Team members that join a project because of a specialized skill or role may retain other responsibilities with other projects and teams. This may create conflicts in priorities and potentially conflicts in roles and responsibilities within the team due to varying expectations held by team members regarding levels of commitment or task responsibility. Strong trust relationships allow team members and leaders to effectively negotiate team responsibilities and levels of commitment, as well as task assignments and member activities. It is healthy and strong trust relationships that allow the team to quickly renegotiate and resolve real and potential conflicts and move rapidly through the team building phases to arrive at strong, high performing teams.

In this environment, some team members may be capable of joining a team, and prescribing trust quickly. This means that team members have assumed trust to be the most logical course of action to take based on the available information, rather than on personal experience. Assuming a trusting relationship is often called swift trust (Jarvenpaa & Leidner, 1998). Swift trust, or the ability to join a team, establish a trust based relationship with team members, and pull together in a cohesive and organized fashion when speed is of the essence, is often enhanced through contracts providing stability to the relationship. This type of relationship is often supported through contracts by formalizing the rules of engagement by which the team will function through the use of documented practices and responsibilities within the contract. Additionally, swift trust is often dependent upon the assumption that a future relationship may be available based on the outcome or success of the current engagement. This too is effectively supported through the establishment of a long term partnership or perhaps a series of potential contracts defined as part of the current contractual relationship. Such stabilizing agreements are often effective in supporting team trust needs.

Swift trust, however, cannot be counted on to provide the kind of trust that enhances long term organizational capability. It works well when pulling key resources and talents into your project from outside consulting firms; not least because consulting firms that specialize in team augmentation or project outsourcing often have the kind of virtual team experience that sets the stage for swift trust to form. Staff augmentation specialists will often seek out long term relationships with companies by developing skills clients require, thus creating within the organization a positive expectation that you will return again when the need arises. These team members join the project knowing they are not expected to stay on the team, but will leave, join another team, and return to your team when you need them. Full time employees, who are more used to semi-permanent relationships, may find establishing swift trust more challenging in situations when the relationships are loose and changing.

The Seeds of Trust

In the past, as work teams came together in the typical, traditional project environment of co-located teams, team members had the opportunity to meet periodically during the day; brief talks around the coffee machine allowed camaraderie to build. Team members could establish rapport informally and develop a relationship that promotes self-disclosure and the sharing of personal information. In this way the seeds of relationships were sown, and trust could grow on a one to one basis. Team members in this environment were able to look at one another as they learned their role, and discovered their similarities and differences. Conversations around the coffee pot, or perhaps the microwave oven, allowed for both visual and verbal feedback to reinforce first impressions. They had the opportunity to reflect upon their team member's physical, as well as verbal reaction to a conversation. During the conversation participants had the chance to adjust their presentation based on immediate visual cues, or perhaps even adjust their behavior or personal appearance as desired based on feedback allowing team members to build rapport over time. The seeds of trust in this way often require constant care and nourishment to flourish and grow.

The relationship building process in the virtual environment is often accelerated by the business necessity that prompted the relationship. Team members may be thrust into their team with their project role often defined by their contract. These teams may have little contact outside of the electronic communications established for the project such as chat rooms, IM, and email.

Some teams rely upon automated reports of progress, and daily assignments as their primary means of communication. Communication entirely through the use of electronic media, by eliminating the human interaction that allows for the natural feedback of face to face relationships, can if one is not careful create the illusion of communication, when in fact communication has not taken place.

Communication and Trust

How do we define successful communication? Remember your first 100 level college class in communications? At that time your professor probably defined the goal of communication as developing a shared understanding of the information. This means that both the sender and the receiver of the information have reached the same, or at least synonymous, conclusions regarding the meaning of the information. Information flow should contain a path that supports the delivery of the initial information and feedback for the information to be truly communicated. If we accept the basic goal of communication as understanding, then information flow is essentially bi-directional. There must be a sender and a receiver, and there must be reciprocation that includes an indication that what was received was understood for communication to have happened.

When a message is necessary, such as the delegation of activities or tasks, the sender of the message chooses what they believe to be the best channel of communication. Often we believe as managers that we have communicated, and yet we only send information in one direction. Effective messaging requires the sender to select a channel, or means of message delivery, that fully supports developing a shared understanding. Channels may be verbal, non-verbal, or written in order to properly prepare the message for delivery.

Verbal messaging may be either face to face, radio, television, or perhaps a webinar or video conference bridge when the audience is large. Non-verbal messaging includes channels such as body language, posture, how we act in a situation, the clothes we wear, the scent we choose, or the environment we create in our work space. Written messaging may be in the form of chat rooms, IM, email, snail mail, automated reporting, and text message to name those means that seem to be the most prevalent. Each of the primary communication channels may be effective when properly chosen and implemented; however in a virtual setting the non-verbal messaging is very often muted at best, and eliminated at worst.

Project team communication in the past was overwhelmingly a synchronous exercise that blurred the role of sender and receiver. Both parties in a face to face communication event were often continuously engaged in communication by simultaneously acting and reacting, thus providing immediate feedback through non-verbal means of eye contact and body language regarding the effectiveness of the message. As teams moved to a virtual setting, the primary means of communication has become asynchronous in nature utilizing written communication as the primary messaging channel. While written communication has become essentially instantaneous through the use of internet based channels such as email, chat rooms, and IM, it has remained asynchronous with a defined sender and receiver acting and reacting independently. While message delivery in the virtual setting is often instantaneous, it may also create periods of silence as priorities remove the responder from the message flow and possible geographic separation of teams expand across time zones.

Figure 2.1   Uni-directional information flow

Geographic separation alone may create periods of silence that stretch from hours to days, or even weeks at a time depending upon the timing of the message. With the addition of geographic separation, often this includes a greater probability of multiple cultures. Adding multicultural teams to the mix may include unexpected gaps in communication due to shifting work schedules, or perhaps even conflicting work schedules due to culturally specific holidays. Teams must learn to adapt their communication skills to avoid, or at minimum account for gaps in the flow of communication.

As teams form a virtual environment the time line to get to know one another is often compressed. Team members don't have the luxury of getting to know one another during breaks in the day, and therefore may not get to know one another's likes and dislikes, schedules, or family habits. With the lack of relationship building may come new challenges in the communication process. Silences may be misinterpreted due to differences in cultural habits. What may be considered a long silence to someone in Philadelphia, may be perceived as perfectly reasonable to a team member elsewhere in the world. Perceived long periods of silence may cause trust to wilt, and team members to fill in the information gaps on their own. Periods of silence, and a lack of response, or uneven communications may be interpreted wrongly by team members that are working remotely from the rest of the team. Gaps in the information stream may cause remote team members to speculate on why the gap exists, and to create a narrative that, rightly or wrongly, fills the gap and damages trust.

Communication and the Team Building Process

Managers need to account for the new challenges in the virtual team communication that may affect the team building process. As teams form, we need to create the time and means by which team members may talk, break the ice, and seek reciprocation and feedback in the discussion. Managers may need to establish a formalized means by which teams come together in an asynchronous format for information discovery. As teams move through the stages of team development the information needs change. Teams in the formation stage require information that supports the development of team norms. By developing norms such as expectations around channels and pace of communication team members can adapt to their new team much the same as collocated teams.

Once effective communication norms are established, the virtual team will be better able to negotiate through the storming phase as they sort out their individual roles, and degrees of freedom to act independently and autonomously. All too often in a virtual setting the level of information sharing is left to contractual agreements as a means of navigating the cultural differences between work groups. We need to open ourselves to a healthy communication process.

What Is a Solid Trust Model?

The theory of trust is examined using three base points in the construction of a collective model to reflect trust in relation to the organization, project team, and team members. Cognitive based trust, personality based trust, and institutional based trust, combine to create a collective measure of trust as presented by

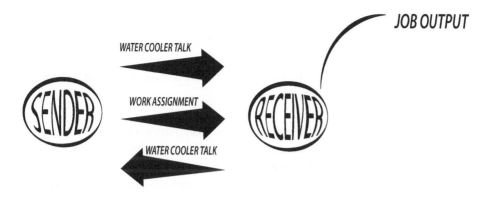

Figure 2.2   Framework of effective communication

Sarker, Valacich, and Sarker in their 2003 study. Trust may take on several forms, as a belief, a decision, an action, and, as indicated in early studies, may vary in degree by gender; however from the perspective of the co-located and global information (IT) team is wholly dependent upon the interaction of three primary base points. These base points may be collectively defined as the three distinct areas of personality based trust, or trust developed during the earliest beginning of personality development, institutional based trust, or that which is dependent upon the degree to which corporate environment, procedures, and policies, may be consistently and cooperatively applied and depended upon as equitable, and cognitive based trust, or that trust which we choose to prescribe based on past, current, or perceived future engagement.

Personality Based Trust is the Foundation

The first layer, relationships, or personality based trust, is the trust which our mothers taught us. "Share with your friends" our moms would tell us. Remember when mom would tell you, if you don't have anything nice to say, then don't say anything at all. This was good for opening doors, and building friendships, but not necessarily good for building trust. Personality based trust has long been described as a developmental reaction to the care provided by the caregiver in the earliest days of life in which the formation of trust through experience is established. As we experience consistency in care giving, and equity in our relationships with parents and siblings, a personality is developed that describes an internal identification of what trust means. In the corporate setting, an individual's personality will be expressed in relationships, and the degree to which the team member may be willing or even able to express trust. Trust is situational, therefore, in nature, and may in part be conditional to specifically related parties, such as within one's own level of trustworthiness, or may in part be related to the individual's own personality traits regarding suspicion of the actions of others.

Personality based trust may affect the acceptance of certain types of communication media, such as in the case of written messages. In early studies, researchers discovered that those individuals with higher levels of personality based trust are much more likely to accept as truth the written word. Where trust exists, so does acceptance of individuals to the team, and therefore openness of expression strengthening and promoting the opportunity for vulnerability and self-disclosure. Maintaining a level of perceived positive reciprocation of communication and self-disclosure is necessary to maintain the level of closeness obtained in the relationship.

Relationship trust is built on open communications that allow people to relate to one another about everything from fantasy football, their favorite online games, Facebook, and work life struggles or family illnesses. It is the shared reality of community that builds up trust on a daily basis, and it is often overlooked by managers and leaders in the work community. It is building in the one to one and one to many relationships that set a foundation for trust across the vast virtual expanse that can build up across the cubicle hallways.

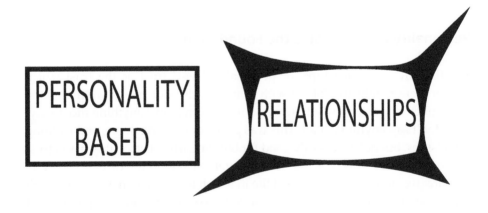

Figure 2.3   Personality based trust as the first layer

Rotter, in 1971, told us that for trust to exist there must be a general expectation of fulfillment, meaning that even when a behavior is threatening it must be carried out for trust to be present. The fear of honest communication with peers, superiors, or employees often stops the foundational layer of personality based trust from growing thick and strong. No one likes to deliver bad news, and many of us hesitate to expose our feelings to peers, superiors, or employees in fear of exposing a flaw that, in the end, really simply makes us human.

In the world of knowledge workers this fear may take on an even more threatening form. It is often the fear of someone discovering that we don't know everything, for in the world of knowledge based work it is knowledge that establishes our value in the work place. Knowledge is not just power, it is the very asset we bring to the company and to those very peers, superiors, and employees with whom we need to establish a trusting relationship.

Every team is established for one purpose, and that purpose is high performance. Very few people are ever hired for the purpose of building friendships, and few people are working because work is their hobby. There is always someone in every work place that will tell you they don't need to be there, but the reality of the situation is, if they don't get paid, they won't stay. Work is not their hobby. Performance is therefore the goal of every team. Whetten and Cameron have described performance as the product of the team member's ability and their motivation, and that motivation is a product of the team member's desire and commitment (1995). Commitment is therefore essential to the effectiveness of team performance. Commitment may, however, be both positively and negatively impacted by communication, identification of one's self as a team member, self-disclosure, and self-categorization, and therefore affect the performance of the team.

So what are self-categorization, team membership, and self-disclosure? These terms refer to how an individual may view their own place in a group or work team. Everyone, at one time or other, will look around the work place and seek to understand how and where they fit. Everyone asks the question "Am I a valued member of this group?" "Am I a member of this work family?" asks the question of whether or not the team member is connected to the other team members, or naturally within the protected group (Halgin, 2009). Connectedness is a term then that answers the question for an individual of whether they identify themselves as a member of the group. It is a sense that an individual senses their belonging, and identifies self with the group when they think about who they really are.

The question for managers is, can we develop a group membership by which peers, superiors, and employees may define themselves as a member of the work team or group? How do we define a group in which membership and connectedness is valued? Communications, reciprocation, and disclosure of oneself to other team members can build an effective foundational layer of trust that allows for a connectedness that draws team members to identify self with the group, establishing a strong relationship that builds trust. This trust establishes an open channel of communication that allows both good and bad messages to flow, and creates the basis for laying down the remaining layers of a solid trust base.

Cognitive Based Trust Depends Upon Information Access

It is hard to find work teams these days that are not, to some extent, acting as virtual teams. Teams are commonly defined as a group of individuals that come together to accomplish a specified task or series of activities that are linked by common goals. Teams will normally work their way through a series of events and problems, and will be challenged to find sometimes unique and often creative ways to resolve these problems in order to maintain progress and stay on schedule within defined constraints. When we add to this the idea of virtuality, we need to understand the unique challenges that virtuality may entail.

Virtuality can be defined as giving the image, likeness, or simulation of the reality of the object, situation, or environment. In this case, the term virtual would then imply that a virtual team is in many ways being given the appearance of a team, or in some way simulating the actions of a team. If we add to this definition the word project, which by definition implies a set beginning and end, with the understanding that virtuality implies a likeness of the reality, then this group of individuals forming a virtual team may be said to have accepted the challenge of accomplishing a set of related or like goals, within a generally defined time and set of constraints, while being provided an environment that somewhat approximates the environment normally afforded a team.

If we accept that virtuality creates the likeness or simulation of the reality, then virtual teams may be expected to be challenged by situations such as having the likeness or simulation of clarity in their roles, norms, constraints, and processes, as well as somewhat undefined boundaries. Virtual teams are normally challenged in having some team members separated by geography, and most team members communicating through electronic technologies.

Geographic separation can include different floors in the same building, different buildings in the same campus, or simply a chosen preferred communication style that creates a degree of separation in the same office space.

When we talk about cognitive based trust, we are talking about the act of coming to know a situation, person, or group to such a degree that we believe we have enough information to make a decision to trust, or not trust. Cognition may commonly be defined as the act of knowing, or the process of coming to know. In the case of virtual team environments, as a result of separation, either through geographic challenges, or simply through the way in which team members may act, the process of coming to know the team and team members is often hampered. Because of this separation even collocated teams that work in a virtual manner tend to rely on cognitive trust, and may be highly virtual and dependent on mediated communication. It is perhaps the reliance on electronic means that cause the team to seek information by which a decision may be made to trust, rather than relying on the personality based processes such as relationships.

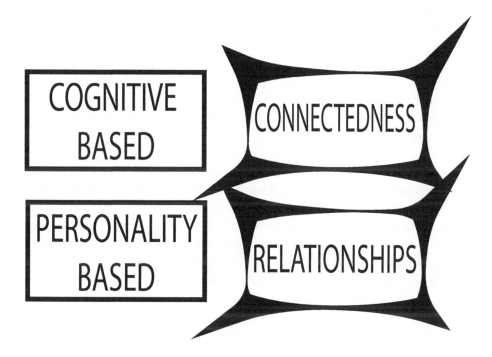

Figure 2.4   Cognitive based trust as a second layer

In order to establish cognitive based trust, team members need information about their co-workers. They need to be able to make sense of this information both in order to assess whether their colleagues are trustworthy and to form an expectation of the nature of their future relationships. In the traditional work place rapport and communication was established informally and over time. In the new (and virtual) work environment, where there is little or no opportunity for old-fashioned rapport building, the absence or poor quality of communication is often cited as the first thing managers should improve.

Team members learn to trust one another as they come together to work. As they reach for common goals they form interdependencies and begin to establish their membership, or *family*, in the group (Fabiansson, 2007). Family is often described as that sense of belonging and common history that forms a lasting bond capable of surviving the challenges and problems that teams may face. It is a sense of being connected by a common vision and sense of purpose. A sense of family, or connectedness, provides for the expectation of a shared future in their membership, lowers the sense of vulnerability in the group, and opens channels for self-disclosure. Thus, having a sense of family, of being connected and coming together to face adversity with the members of the team, provides for greater opportunity in self-disclosure that offers opportunity to enhance the gathering of information such that a decision may be made to trust the members of the team. As team members work through adversity, they come to know one another's strengths and weaknesses, and build expectations regarding each other's capabilities and modes of operation. Over time, team members are able to build a strong and healthy reputation that may precede future work engagements.

Reputation development within the team may further enhance the feeling of being connected to the group. Over time, members may strengthen their role and connection with the group through daily activities and the process of reinforcing expectations of themselves. The day to day give and take of working on a team has a way of solidifying one's place within a group by a continuous process of knowledge development. Reciprocation, or give and take, is the process of interaction that team members use when providing personal information.

A final note in cognitive based trust development is the managing of stereotypes. Stereotyping occurs much more often than any of us would like to believe. These stereotypes must be managed on a daily basis by team members to avoid team member assumptions being formed on the way one is dressed, how they speak, or some other trait. We all must manage our impressions of one another on an active basis to affect an image that we are all consistent and rational as individuals and capable of fulfilling our team role. As teams form in a virtual environment the process and time to build trust among team members is shortened, and may be highly dependent upon one's reputation. Team members are often confined by circumstance to depend on our past experience of team members' work, or the experience of others within the organization with an individual or group as a means of establishing trust. The establishment of trust may, in a virtual team, be initially dependent upon the generic stereotypes developed over time regarding a group, and therefore non-stereotypical behavior may undermine the development of trust.

Institutional Based Trust is a Corporate Issue

We, as managers, often seem to think that employee behavior is driven by the rules, values, and generally accepted interactions of the organization in which they work. One point that is often misunderstood is the idea that what happens in the office always has three distinct perspectives. What the company thinks is happening is the first perspective, what the managers believe is happening is the second perspective, and what the employees are really doing to get work done is the third perspective.

What the company wants to happen is often driven by the strategic planning process. Within the strategic planning process should be the set of human resource policies that include elements of change management to ensure that what managers believe workers are doing, and how work actually gets done, aligns with what the company wants to accomplish. As managers we build processes, guidelines, and work instructions that we intend to lead the way in which people accomplish their work. This is intended, for the most part, to ensure that work is repeatable and measurable in hopes of ensuring that work practices are the same across projects and independent work groups. By providing these guidelines, we as managers hope to make our programs predictable.

At the level in which work is actually accomplished, employees realize that the actual practice of getting work done is a continuous negotiation of very specific daily priorities. This includes the negotiation for basic scarce resources such as printer time, server time, project team members, and the attention of individuals that may be considered subject matter experts (SMEs) to identify and resolve the more difficult aspects of the daily just get it done priorities. As a result of this, what may happen at any given moment may not match what the managers and companies believe will happen to accomplish a task. This problem can have a strong tendency to affect the perceptions of equity across work groups, projects, and individual team members in the application of organizational procedures.

As individuals compete in the negotiation process, or perhaps simply in the accomplishment of their daily project tasks, issues of equity may impact the level of conflict within a team. As team members look around, and perhaps find that they are falling behind in the project while adhering to the expectations set by managers regarding the performance of their work according to the prescribed procedures, and perhaps other teams or team members are getting ahead while bending these rules, questions of equality in practice may arise. Perhaps as teams or team members are rewarded for bending the rules in order to remain on schedule, and other teams or team members may be punished for the same practice, this too may raise questions regarding equal application of the rules. As time pressures build on a project, it is often the practice for teams to work overtime and weekends to make up for slips in the project schedule. Differences that may arise in work practices, such as the differences in cultural attributes that include national holidays, religious observation of holy days, length of the work days, and practices regarding leadership and conflict resolution, may have an effect on teams as they work extra hours and days to accomplish the project. One of the major challenges in teams that work virtually is the higher likelihood that these teams will consist of multiple cultures with differing levels of expectation regarding adherence to procedures and guidelines.

Predictability, or the degree to which leaders consistently reward behaviors that support the achievement of goals, and the degree to which desired values and norms are reinforced, enhance institutional trust (Gillespie & Mann, 2004). This means that predictability and reinforcement of expectations does not always need to be a positive experience, and that maintaining the expected corporate order is a large part of predictability. Just as our parents taught us, and we have taught those around us, good behavior needs to be reinforced, and bad behavior needs to be extinguished. These simple rules maintain equity, and build institutional trust.

In the situation of cross-cultural teams the process of feedback and reinforcement may be confounded by differences in culture or organizational

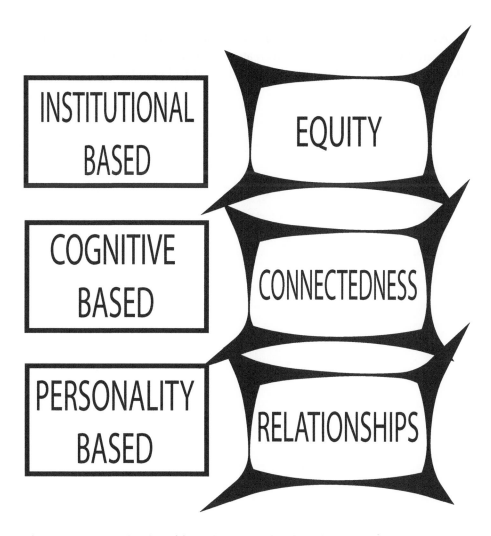

Figure 2.5   Institutional based trust as the third layer

position, as in many cases in which the collective nature of the culture may hinder the process of negative feedback (Garza & Lipton, 1978). Collocated team members' development of trust is also significantly identified with perceptions regarding equitable decision making, and may additionally have an effect on pride and respect further affecting the processes of vulnerability and self-disclosure necessary for the first two layers of trust to develop and strengthen (Lipponen, Olkkonen & Myyry, 2004).

As you continue to read, I think we are now ready to develop solid, real life means by which a manager can create a positive environment for trust, and effective practices by which trust is strengthened and maintained.

3

Building Personality Based Trust

Personality based trust is built upon the relationship skills our parents taught us early in life. As young children we build a framework of what, to each individual, trustworthiness means. Trustworthiness, openness, and a willingness to reciprocate, as well as some aspects of our attitude to risk are traits that each employee brings to the work place and are the tools upon which we can build. It is up to the management team to create a work environment that supports personality based trust. How can we, as managers, build such an environment? Won't employees naturally sort this level of trust out for themselves? The answer is both yes and no. We might simply assume that the employee's natural instinct to build relationships will take care of this problem without the need for any active management. Relationships will form based on geographic, ethnic or socio-economic as well as psychological empathy. This will of course happen regardless of the environment we build, but it does not happen well on its own, nor should it be simply an accidental process.

When I first began working for the local power company as a very young man I worked the night shift shoveling the coal back on the conveyor belt that fell from the conveyors during the day, and washing down the conveyor chutes to prevent dust explosions. Each night we would drag ourselves through the same grimy routine alone in a dark conveyor chute covered in greasy coal dust and sweat. We would stoically plod our way through the night to complete the task, and when done, we silently reported in to our boss and went home. We worked for our pay, and nothing more.

We never met anyone else that worked in the plant, and we really didn't care much about the outcome of our work. There was no attachment to the company or to other workers, nor did we have any idea how our work affected those who came into work after we left. We were not members of any work team, and our only allegiance was to our paycheck. One day the morning supervisor came in early to totally chew us out. He didn't hold back, screaming about how every day his shift started up late because of the mess we left at the bottom of the chute. Really? It seemed to me that at the bottom of the chute was where all my hard work ended up, and yet I had never really made it to the bottom to take a look. The problem was there never was anyone at the bottom of the chute, and so I didn't know what took place there. We nightshift people never saw the face of a co-worker, and I therefore had no reason to care about the bottom of the chute.

What did make me care was when the employee who had to clean up the mess, met me a couple days later when my behavior hadn't changed. A few days later I got to meet the maintenance team as well. They showed up late in the evening one work night to interrupt my routine and make repairs to the conveyors due to the unusual load caused by my routine mess at the bottom of the chute. Now my work life actually had faces and names, and teams and groups, and feedback. I got lots of feedback.

Working in a virtual team setting can be like working a perpetual night shift. You may never have the realization that there are real people working at both ends of your shift that are impacted by what you do, or don't do. The real question we want answered as managers is how to connect the perpetual night shift with the rest of the team. As managers we want our employees to identify themselves as a member of our team, not as disaffected employees with no stake in the company. This begins with an emotional attachment, or better said, as the degree to which someone sees themselves as being connected with, and having a relationship with, other team members that is worth the effort to maintain. This problem of disaffection becomes more complicated in a virtual team environment. The first day in my new job working for a large mass media company provided a great perspective on the problem of trust and relationships in this virtual project world. Fifteen minutes into my new role as Director of Quality Management Systems I took part in a meeting which was basically a dog fight over process improvement failures, and who was to blame.

In order to provide changes to the way in which the engineering process works, teams had been organized across several regions and organizational boundaries. The effort was aimed at reducing those systems defects that escape the test process and became customer facing problems. Over several months frustration had set in at the inability to get any real improvements, or for that matter to come to agreement on how to get the improvement process started. One team's focus was to identify ways to better communicate test requirements for test planning and scheduling to shorten the test cycle time, while others focused on defect reduction, defect triage, better test automation, and coordination issues between widely dispersed engineering groups. As I watched all the rock throwing it became quickly apparent that the majority of the folks in the meeting had never actually met, or even talked at any length with their engineering counterparts in the project team. Test lead engineers finally admitted they didn't even have the names of the development or architectural team members, and had no way to contact them with questions or concerns on the project.

How Do We Fix Breakdowns in Personality Based Trust?

In management science, there are many different styles in which managers may engage employees, make work assignments, and build rapport with their employees. Moore (2007) noted in a doctoral dissertation that choosing a management style that promotes greater opportunities for team members to socialize, getting to know one another and come together as a team are essential elements in team motivation in a virtual setting. Additionally, the extent to which team members are able to meet face to face aids socialization, shared concerns, and team identity, and may provide the greatest advantage in building trust relationships. Socialization and communication are essential elements in building a bond of concern and team identity to build upon their childhood perceptions of who may be trustworthy (Kerr & Kaufman-Gilliland, 1994).

Employees that experience isolation tend to experience lower levels of trust and commitment to the work team, and may often be compelled to make decisions between their own self-interest and that of their team, and may seek out greater isolation from the team (Godar & Ferris, 2004; Kerr & Kaufman-Gilliland, 1994). Our role as managers is to build the environment in which personality based trust may grow strong naturally within the work team settings. Often in the work place, employees tend to gather with friends from their days in school or from leisure time activities, or even relatives that may share the same work place, thus extending their home life relationships into the work place. We, as managers, need to encourage an identity based on our employees' work relationships in order to help them to identify their work life as a valid opportunity for solid relational connections. Managers should build a cadence around team gatherings and communications; reduce the complication of project plans, and create an opportunity for shared leadership.

The keys to virtual team trust may be described as an expectation for future team efforts, better solutions to complexity and short periods of high activity, a predictable cadence in communication, opportunity for sharing of one's own self-disclosure as well as being open to another's, and the sharing of one's past good experiences with the present team. This requires solid planning and an effort on the part of the management team to accomplish all of this, and yet, it seems our children have done very well in this virtual environment. Team socialization is a necessary tool we as managers must provide to address what Maslow[1] notes is necessary for social level actualization before a team can grow into a greater level of satisfaction and motivation. When working to build effective virtual teams a manager relies heavily on understanding the context and situation of the team; for example, employee feelings, problems, work issues, and other conditions affecting motivations and team member interdependencies, in order to demonstrate awareness and empathy for team member needs (Weems-Landingham, 2004). Additionally, the ability to listen attentively and actively in a non-evaluative way when facing the absence of face to face communications is heightened when managers use all available electronic means to monitor employee conditions.

Therefore the manager of virtual work teams must enlist every available means in the new world of electronic communications to listen and gather facts. This allows the manager to fully understand the factual conditions, as well as employ the traditional methods of listening, empathy, teambuilding, and participation. Text messaging, instant messaging (IM), email, and other electronic media can easily become effective strategies for building trust in work teams and should not be a disruptive or limiting agent in communication and team socialization (Wise, 2011a). Managers, in an effort to build effective teams, must find creative ways to create a safe environment that supports personal vulnerability to allow socialization if the team is to be a truly effective virtual workforce (Wise, 2011a).

---

1   In 1948 A.H. Maslow published a paper analyzing the theory of basic needs gratification in complimentary relation to the contemporary motivational theories of the time. This analysis became known as Maslow's Theory of Hierarchical Needs with the basic tenet of the hierarchy being as a basic need becomes satiated, or fulfilled, it is replaced by a new and higher level need. According to Maslow's theory the most basic needs must be met before being replaced by a higher level need. Maslow's hierarchy of needs is often portrayed as a pyramid with the base need described as physiological needs such as food and health, followed by safety, belonging, esteem, with the highest most layer of the pyramid containing the need for self-actualization.

The Context of Social Media

According to Lenhart, in a 2009 study by the New York Department of Health and Mental Hygiene, 77 percent of teens used an online social networking. In only three short years this number has increased to 86 percent based on a 2012 Pew internet study (Brenner, 2012). This number drops to 72 percent for participants between the ages of 30 to 49 (Brenner, 2012). When using these sites 83 percent added comments to a friend's picture, 77 percent posted messages to a friend's page, and 71 percent sent a private message to a friend's social network site (Lenhart, 2009). These statistics are impressive and growing, and these teens are now moving into the workforce in large numbers. Among young adults the numbers are even higher.

Studies have shown that electronic communications can have the effect of lowering social inhibition and encourage participants to disclose more personal information making connections that may not otherwise have occurred. In this context, according to Brenner (2012), in most cases of online communication people report the interactions to be perceived as kind. In other studies it was discovered that people are often better able to access their *true self* when interacting over the internet. With this in mind, and in keeping with the thought that folks working in the virtual setting will often never have the opportunity to talk face to face, an internet coffee bar is a great alternative method to allow time for people to get to know one another. Bargh, McKenna and Fitzsimmons explain that internet connections that result in events of self-disclosure, if repeated over time in the similar internet interactions, may lead quickly into friendship (2002). Self-disclosure can lead quickly to a bonding that includes empathy and understanding between people. By encouraging these bonding opportunities managers may increase the level of trust between team members allowing for greater team interaction and identity, and thus likely increasing team performance.

Teams need a place to bond, and in the absence of the traditional work setting, managers need to provide a place that describes the vision and mission of the organization, the annual goals and achievement, project specific information, and a place that the employees may call home. Why is such an information sharing structure important you may ask? Because it ties, or connects, the employee to the pace and direction, or heartbeat, of the organization. Some of this builds a point for personality based trust, and other parts of the structure are important to cognitive based and institutional based trust, because for the other trust layers, as with personality based trust, people need to be able to discover information in order to build and nurture trust.

Being connected means that an employee now identifies an important part of self as belonging to the organization thus enhancing their ability to trust, lowering their guard, and enabling them to allow others close, enabling self-disclosure. As Figure 3.1 depicts, according to Whetten and Cameron (1995), individual performance is a function of the employee's ability × motivation, and motivation as the product of desire and commitment.

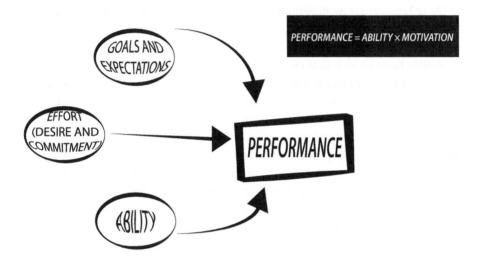

Figure 3.1   Ability and motivation are essential to performance

While there does not appear to be any evidence that personality based trust is affected differently based on the role that a team member may play (Wise, 2012), managers must be aware that each employee carries to the work place their own set of experiences and expectations. As we have learned to this point, commitment is essential to the well being and effective functioning of a team and is tightly coupled with team member commitment, and may be positively or negatively affected by communication, team identification, self-disclosure, and self-categorization, and therefore can affect the overall performance of the team. Finally, a shared identity can have a positive effect on the pride and respect an employee may feel which can in turn have an effect on the process of vulnerability and self-disclosure which may have a positive effect on conflict management (Lipponen et al., 2004).

Establishing the Basis for Virtual Communication and Knowledge Sharing

A short vision and mission statement, as well as products and services, not only serve to inform your customer base and provide an inexpensive way to attract a new generation of customers, but may also serve to inform and update and attract new and current employees. The first question I always ask a prospective employee is around their knowledge of our company's business. Knowing what business we are in is an essential element to belonging, and yet one of the hardest questions to answer for non-C Level employees. This is one of the essential elements of membership, and yet an often neglected point of on-boarding.

Next, team members need to be working toward a common goal that provides opportunity for shared leadership. Common goals provide such an opportunity, and yet, without a place where goals are established and maintained, teams may easily lose focus. Keeping goals updated and upfront helps project teams stay focused on the measurable deliverables by which they themselves are to be measured and awarded. By sharing the leadership needed to reach their goals, buy in and attentiveness are better achieved.

Shared leadership means that as project milestones are met, and new milestones come into focus, the team member, or members, with a direct stake in the milestone outcome step up and take charge. As a team member's expertise becomes the focal point of the project, the team's expectation may be that this team member becomes the subject matter expert (SME), and takes a leadership role in defining how the next milestone is to be accomplished. This provides each team member opportunities to grow, perhaps learn new skills, and have an opportunity to connect with people outside of their immediate span of control and influence. Tracking, reporting, expectations, and roles should be charted and tracked in a centralized place that is reachable, and owned by the team. By doing this, team members are able to establish a presence in the team that allows for well defined roles, clearly defined responsibilities, and a sense of shared activity, all of which may help in bringing about a perception of safety, therefore allowing for a more effective level of self-disclosure (Fabiansson, 2007).

Another simple means of getting started can be a wiki site such as Jive SBS or Sharepoint. Each of these sites may provide a place to gather information such as a searchable wiki for employee pictures and corporately relevant employee biographies and blogs. Peer training information and data sharing are great on these sites as well. Being able to have a presence that is in front of, and readily accessible to, other team members helps every employee to feel like they are heard, relevant, and important to the team and team processes such as learning, conversing, and getting to know each other and the company.

Chat and Instant Messaging Keep Everyone in Touch

Since trust is normally considered to be closely tied to consistency over a period of time, then openness, vulnerability, and self-disclosure may be hampered due to communication problems during team ramp-up. This problem can be even more relevant to "geeks" as they tend to be heads down and involved with their technology more so than most people (Glen, 2003). To knowledge workers such as a software developer the computer can be a closer friend, at times, than the person in the next desk over, so the need to keep a technology based communication tool close at hand is very important.

I know for myself, and while I don't really consider myself to be a geek since my knowledge of technology is not that great, I avoid phones if at all possible. It seems at times that the thought of picking up a phone and asking for anything is intimidating, and yet a quick text message or instant message using IM is fine. I make the effort to get it down in type, and off to the recipient without a problem. The use of electronic communications has become so ingrained in the way we as people operate and reach out to one another that, while it seems not very personable to many folks, it is simply the expected mode of operation.

To be sure that project teams stay in touch without delay, and without hesitation, managers need to be sure that project team members use all available technology, and are capable of reaching out to all other team members in a way that is relevant to them (Wise, 2011a). I often see employees working with several IM windows open and active. These team members, even ad hoc teams that come together with urgency only to accomplish a quick task, often need to work across thousands of miles, and need to work using the simplest and quickest means available to them. This quick means is often instant messenger services with a shared desk top allowing everyone to see the same computer screen. IM can of course be complicated when teams cross time zones, however my team and I tend to share this challenge by planning our meeting times out such that we trade off early and late hours to allow for synchronous communications.

Such openness and availability can build an opportunity for vulnerability and self-disclosure, and may at times be a way to simply signal an openness to, and desire for, working closely with co-workers (Nardi, Whittaker & Bradner, 2000). IM and text messaging also provide a lack of formality that establishes a relaxed tone to the conversation that enables participants to be open and vulnerable in the conversation (Nardi et al., 2000). Informality in a conversation can often open team members, co-workers, and even peers in the management system to participate in problem solving when they would not normally be open to such opportunities.

I have often experienced the difficulty of engaging a peer in problem solving when using the standard phone call. Nowadays the phone call just seems an intrusive interjection of someone else's agenda into a packed schedule full of our own problems and priorities. I asked a co-worker to IM the needed person from finance after several hours of ringing the phone and sending emails to try and raise the concern to the proper level. The amazing thing to me was this was a peer of mine, and a director level person in finance that was absolutely necessary to solving the problem, and yet, when I asked the engineer to IM the director, the problem was resolved in under five minutes of effort. The informality of the communication, and a previously trusted relationship, solved the problem very quickly. As Wilkens and London (2006) noted, the use of closed loop communications can contribute greatly to problem solving, and IM can be a ready and informal closed loop process.

Face to Face on a Cadence is Important

Although we have spent a lot of time talking about how folks prefer electronic communications such as email, face to face time remains an important, necessary, activity in building trust, and needs to be available for team members to interact with peers and team members visually, in a setting where the visual cues are available. Face to face opportunity needs to have a regular schedule, or cadence, that supports the needs of the team in relationship development. As we noted earlier, working in the same office space no longer implies that face to face interactions will take place. Even if folks are good at working together and open to talking face to face, and they already trust one another and are happy to be in each other's presence, in the mobile office world of today's business environment, people simply are not always working in the same work place every day.

As I was writing this paragraph I received an email from our office manager noting that today she would be working out of the New Jersey office instead of the Pennsylvania office, and wanted us to know just in case we wondered where she might be this morning. Employees are fully able to work from home, the local coffee shop, or in the case of one peer of mine, stopping along the side of the highway one day to participate in a web based meeting because traffic was bad and he was unable to reach his office in time. For myself, after experiencing a flat tire on the way to work one day, I spent the better part of the day logged in and working from an interstate rest stop.

So my point is this. Assuming that because people normally attend to their work in the same office location, that this means they regularly take advantage of face to face communications is really not a safe assumption. Many people today simply prefer, or as in the examples above, require the freedom from the traditional means of communication in order to fully participate in the conversation.

It is up to the manager to make sure that face to face communication opportunities are provided on a regular basis to maintain the human factor in the relationship. A common problem some studies have noted that managers may face is the potential that the rate of adoption of technology based communications and the preference for face to face communication opportunities may be demographically influenced. Younger workers are likely to quickly adopt electronic communications and new technologies, while older workers may choose to weight the value of adding the new technology more carefully (Morris & Venkatesh, 2000).

In other words, the younger the crowd, the more likely they are to prefer electronic communication over face to face interactions in which everyone gathers in the same room. Also, the younger the crowd, the more likely they are to prefer ad hoc video teleconferencing over a regularly scheduled meeting (Moore, 2007). Trust, though, is increased when people can meet regularly face to face, even when the face to face opportunities are through electronic means.

Using electronic means to mediate face to face communications can, and very likely will, have a positive impact on personality based trust (Karpiscak, 2007). According to Karpiscak, if a team has even just one face to face interaction, then trust is better supported (2007). This may be true for the same reason that my experiences shoveling coal on the night shift at the power plant changed. Meeting face to face can make the previously unseen virtual co-workers real.

I'll give you another example. The greater part of my graduate work was done online. In two years of my education I had never seen anyone else in the school. Not one other learner, and not a single instructor had ever been met, or even talked with beyond email or chat groups in the course room, and I had to admit, during my very first face to face meeting, that I had begun to see the entire organization as a black box. Now, this was not the fault of the school or any of the other learners in my classes. As said earlier, a virtual team mate may seek to remain unseen, and may seek out greater isolation from the team. As a result of my own actions, I had begun to wonder if there really was a real person on the other side of the online course room.

Now the delay in meeting people was in large part my own doing. I had avoided all of the necessary face to face opportunities that the school provided, but again geeks and young people tend to do that. OK, so maybe I do have a level of geek in me, but I like to think maybe it is just that I have something in common with the young, up and coming, new generation of workers, kind of? After only one meeting face to face with a real instructor I was bought in anew. I now had a face matched up with a name and a personality, and I understood once again that there were real people on the other side of the deadlines and commitments. While I had previously developed a small level of trust as a result of the interactions that I had using written communications, my level of trust and respect increased greatly following this meeting.

Surprisingly, I also began to find other learners in the same program after this meeting. I think that this was due to a new, heightened, level of trust that opened me up to being open to discussion and self-disclosure about the education program. Once I began to speak more openly about the program I found that others were willing to reciprocate and draw closer due to their shared experiences. Most amazingly, I even found that a friend of mine was a learner in the same program.

Some companies, as in the case of my educational experience, fly team members about the world to ensure they meet in person, face to face, at least once in the beginning of a project. This breaks down the barriers, opens up communication, and allows the team to bond much quicker. I have done this in my current role when I believe the need to bond is critical. Recently I took a quick trip to Chicago to meet with some folks in the operations division of my employer to give a face to my department. Simply by being in the same room with the folks in Chicago for the first time, we were able to break down barriers between our groups that had lasted for years. We talked about the problems our groups faced together, and ways to begin working closely to resolve differences in practices that were causing us problems. We now have a relationship, where in the past we simply had an email presence.

Video Conferencing, iPhone—Facetime, and Skype

In order to meet face to face, all participants do not have to be in the same room. There are many good tools that we all have available for creating a face to face experience. My employer has begun investing in iPhones, Skype accounts, and video conferencing equipment. These tools represent the range of available tools in price and capability. Of course there are many other tools that can be provided, but I wanted to at least share the ones that we have available to us. Skype, ooVoo, or other web based video phone tools are inexpensive, and can be implemented with relative ease. Get an inexpensive web camera and the team is ready to go. We keep these cameras around and available so that teams that work together often have at least one camera available for meetings at all times. Seeing, and being able to pick up the social cues involved, help greatly in understanding a conversation.

Another great tool that is readily available is the use of an iPhone, or other video capable smart phone. We have recently begun using the smart phones instead of the standard email enabled business phones. With the smart phone, such as the one Apple has made famous, an immediate video conference is at your finger tips, or better said, in the palm of your hand. Hit the button, and your team mate is available by video phone and becomes real before your eyes. You can immediately pick up the cues.

Are they stressed out? You won't necessarily pick that up in a text message that says, "I won't make my deadline." I'll never forget the day in 1990 when I got a phone call from a youth league leader following a big storm. He called my home and said, "Oh good, you guys are fine. I thought I kept you on the phone too long while the storm was coming," he said. "I'm glad it missed you," he exclaimed. What he could not see was the only thing left of the home in which we answered the phone, was the phone. The entire house was gone around us, but the phone, to our amazement and somewhat ironic amusement, was still working.

The person on the other end didn't quite get the full impact of why we were a little short that day on the phone, but the stress was too big to express at the time. I was not going to meet the deadline that day. A video phone, though, would have made my needs immediately apparent. To a lesser extent a simple tool such as a smart phone, for teams that work so far apart, can be a very inexpensive way of making these connections, and building strong, solid, and effective relationships, and opening the doors for trust, and effective problem solving.

Music and Game Servers, and Shared Down Time

In the old days, and sometimes it is still the way to go, team members would take to the links for a nine hole round of golf to fully bake their ideas and at the same time build a stronger working relationship. Co-workers still rely on company sponsored softball teams, bowling leagues, and many other team sports around the world. A problem now arises when co-workers are spread around the globe physically, yet connect at the speed of fiber optics and satellites. So how can a manager connect co-workers and team members in such a way that they get the benefit of competing as a team, and thus building team cohesion? Teens and pre-teens all over the world do this every day through Xbox, Wii, PlayStation, and online game servers that support team play.

When I worked in the financial world some information technology (IT) guys received permission to put together a game server that they could use before work, during their lunch break, and after hours. The game gathered together players from all over the company. Teams were formed, and events planned that lasted hours into the evening. Players created an online presence with personal banners. Game referees kept stats on players, and players were able to build a reputation for themselves that they paraded around the game server proudly displaying their achievements.

It was online, but at the same time it followed them into the work place the very same way that physical team sports follow participants. Players received the complimentary comments at the beginning of meetings from participants and non-participants alike, and their gamer tag was often used when communicating with others on team and project issues. Players wore their "game jersey" by using their personal banner. This simple opportunity to participate online with other workers whom they may have met only through email, now provides a strong link that would bring them together emotionally.

Rules for use were established and approved by the company, and a simple one time investment was provided by giving the responsible manager a used and decommissioned server. Access stats were tracked to be sure that the participants did not take advantage of the privilege, and the system was monitored to ensure it did not provide an unauthorized access path into the company network. According to Ducheneaut, Yee, Nickell, and Moore (2006), games played online with players that have never actually physically met can form tight groups that support the feeling of belonging. Online games offer the opportunity to belong in an atmosphere where other means may not be present (Wan & Chiou, 2006).

Remember that being connected is very important to feeling connected, and therefore creating the sense of belonging to the group. Yes, obvious, maybe, but important. This feeling is enhanced by the employee's position within a group. If an employee is considered to be lesser to those with whom they work, either in years of employment, skill, stature of some sort or another, or in a position of less importance to the group, then they are less likely to feel a sense of belonging (Weyuker, Ostrand, Brophy & Prasad, 2000). Online gaming provides the sense of participation on a level field that may just be what it takes to bring together team members in a way that supports equality. The games allow for the activity of gaining in stature and achievement among peers as well as an opportunity to build relationships with those they have no other opportunity to meet (Yee, 2006). This forms an opportunity for a degree of equality where it may not otherwise be found, and establishes the chance for a degree of vulnerability to develop, and therefore the foundation for a trusting relationship to be laid. Wright, Boria, and Breidenbach (2002) noted that the weak players are brought into the group through supportive behaviors by the more experienced. This gaming environment may offer those on the fringes of the group, either in skill or stature, to come closer to the group, and safely experiment with their position within the group culture. Games may also allow for the culture of the group to grow and strengthen while strengthening group norms and expected behaviors.

4

Building Cognitive Based Trust

People are thinking beings. We never stop using our mind to interpret the world around us. We take in information about our surroundings and filter out those things we find irrelevant, sort and categorize what we find important, and make decisions as we go through our day. When we find that gaps exist we fill in those gaps with whatever information we can find. In many cases, in the absence of direct, relevant, and sensible information, we fill in the gaps with our imagination and our fears. The point is cognitive trust is all about making choices, and information gaps may open us to misinterpretation, misinformation, or rumor.

Rumor mills never stop. When I worked in the electric power production industry we used the rumor mill to our advantage at every opportunity. Rumors were created to keep ourselves entertained. One of the ways we had the most fun was to start a rumor just to see how long it would take to get back to us. It never took long for a rumor to make the full cycle and return in the form of a wild eyed, panic stricken co-worker winding a drama filled path through the sea of cubicles to frantically exclaim the latest rumor scoop. Sometimes if we really needed something to change we would start a rumor, and a few days later a manager would come by with a great idea. Curiously, the new idea would often be suspiciously close to the latest rumor we released into the crowd.

Gaps in the information flow can be particularly difficult for remote team members to handle. The gaps can easily be interpreted in many ways by team members such as disinterest, lack of activity, lack of belonging, or as a lack in desire to share information with remote members. A decision to form offshore or virtual team relationships should include a willingness to share information and needed corporate knowledge.

Often, when focused on the work setting, and perhaps more prevalent in some cultural groups, in order to make a decision to trust someone, information needs to be available. Curry and Fisher noted in a 2012 study that familiarity can be a necessary precondition when deciding to trust, or mistrust, another person, and perhaps can be even more important than the warrant of another person regarding trustworthiness. The old saying that familiarity breeds contempt, to some extent, may have some truth to it. As we become more familiar with people with whom we work, we become familiar with their ways. These interactions and experiences build an expectation for future behaviors allowing team members to make up their mind whether or not to trust another team member.

Team members working in the virtual model often find it difficult to become familiar with their team to such a degree that the relationship has moved beyond the introductory and discovery point, and into a trusted relationship. It can be difficult to discover information about people working in the same office, and almost impossible to become familiar with everyone on a project. I was recently embarrassed as I began a new teaching assignment. Each time I begin a new semester the class, of course, begins with the introductions of myself with the new classmates. As I have explained, discovering information leads us to trust one another, and the classroom environment is no different. I introduced myself to the class, talking about my past jobs and work experiences, education, and my current role and job location. As the opportunity moved from learner to learner, each in turn introduced their roles, jobs, and education. Finally, one person in the class, with a big smile, exclaimed to the class how she too worked in the same office as myself. We had never met, yet worked only desks from each other every day.

She always entered, she said, from the back door, and had never walked to the front entrance of the building. Those of us, and yes that really is most of us, working in a virtual model often have to rely on a decision to trust our team mates because we lack the personal interactions that allow relationships to grow. We have to decide to trust others based on information we can find. This information is often found in one of three main categories called *unit grouping*, *reputation*, and *stereotyping* (Sarker et al., 2003).

Unit Grouping—Have an Information Plan

Unit grouping is the process involved when team members come together to accomplish common goals. They form interdependencies and a sense of membership within the group, or otherwise known as that home feeling, and a sense of safety in the group with a shared identity. Feeling connected, as the old saying goes, may be very closely related to the effectiveness of the communication. So, how is communication measured for effectiveness? Is the purpose clear, concise, and relevant to the reader? These are the necessary elements to the first part of the question. The next part is the more difficult question. Is it consistently delivered? If we get this part right we can reduce the risky behaviors an employee sometimes follows when information flow is filled with gaps. We can also increase their sense of belonging.

Information, even delivering a bad or unwanted message, is important to developing trust. Sometimes knowing that someone cannot be trusted develops trust. We simply learn that we can trust someone or something to not be trustworthy, such as knowing that a group or person will always be late, or deliver poor quality, creates trust through the consistent delivery of a bad message.

As managers we need to create an information plan. What do we want to get done, and what do we want to affect, are the key questions to ask when building this plan. The information plan needs to have an expected outcome to be effective in sending a message. This is the clear and concise part of the message. Remember the old adage, what gets measured, gets done? Getting work done, regardless of the need for trust, is every manager's job, so we should then be focused primarily in the area of getting work done.

The next thing to remember when developing your information plan is the part about team make up, or relevancy. Project teams, or groups of employees, are not a homogeneous group. They come from different corporate divisions, different divisional departments, and in the case of the modern project swarming effect, often different companies with many divisions and departments, and in many different geographic regions and countries. This requires a plan that is specific to the project or goal.

Let me tell another quick story, and then we can get to the point. The department in which I work spent over a year planning just such a plan, and creating a wonderful *management dashboard,* to relay every type of information a project team member could possibly want to gather. It is great! Every piece of data is available to anyone that cares to look. Personnel hours on every project are collected and displayed down to the individual participant's time commitment. Defect detection rates, defects that are missed and passed to the customer, and so on, and so on, and so on, and then comes the question of, "So what?"

Dashboarding Builds Unity

In keeping with the new trend, dashboards are required, and comprehensive, however, when the question of relevancy and conciseness are placed in the equation, the so-what question, the dashboard can tend to fail. In order to get to the relevant part, the user must click through a layer of imbedded drop down lists. OK, so not bad right? Unfortunately the drop down lists are almost always non-descriptive names and numbers designed to guide the knowledgeable seeker, or other knowledgeable person in database design, and not the remote user who is simply trying to stay connected. After over a year of development, very few people have ever cared to use the new dashboard, and those who do get misinformation because they are not sure what the information is trying to tell them. To avoid this problem we need a plan. Create a plan based on what the organization wants to affect, and build from that point out.

The plan should be layered. How do interested parties find information about the business, and about the project, and about the people? Remember that information discovery is the key to cognitive based trust. Using a drill down method for discovery provides some context to each of the different layers. Begin with taking a good long look at the corporate or division level goals. In general most of the goals will fall into four categories of Quality, Effectiveness, Efficiency, and Productivity (Arney, 2011). A few of you may see these translated a little differently, but for the most part these are the primary goals with which we all deal. These are so common that anyone reading may immediately feel like I stole this from you. Maybe! All of us have begun at this point in one form or another. See Figure 4.1.

Now, once again, let's back up a little to talk about why we are starting this conversation at such a high level. Cognitive based trust begins with the ability to discover information. The discovery is important at both an individual level and at the group level. Individuals need the ability to discover information about each other's personality, likes, dislikes, family, and friends, however, if we followed many of the suggestions in the last chapter, we should be doing pretty well in this area. At the same time individuals need to be able to discover information about groups as well.

Groups such as work groups and project teams must also be discoverable. It is important because it is at this level that organizations are best at identifying, categorizing, and reporting project and department level data. To be successful we need to do things in a way that make sense for a department or company,

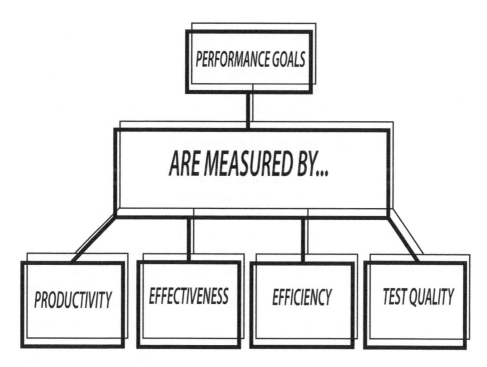

**Figure 4.1   Example communication strategy chart forms the communication
          strategy**

and can be made sustainable and repeatable. Keeping the data collection at
this level will allow all of us to make these changes and implement them
successfully.

Additionally, reports at this level allow individuals to seek out and discover
information about dependent projects without needing to know about all of the
individuals involved. Team members can now discover information about a
project with which their work lives intertwine, and build the ability for cognitive
based trust, based on the reporting. This person can take the information
and begin to drill down, seek out consistencies or risks, and build enough
information on which to make a decision to trust. This type of information is
important to teams that have dependencies, but little connection. Many times,
as a result of divisional silos, dependent projects will never connect beyond the
high level project management office (PMO). We can all relate at least to one
catastrophe that may have been avoided if we all had a functional reporting
system.

I'll give my own quick catastrophe example, and then we can drill down to the next level. I had just begun a new role as quality manager and was attending one of my first project meetings. The project, and this is the perfect moniker for every oversized project, was given the name *the perfect storm*. The project started around the time of the movie release, and it truly was perfect. The name, real or not, is perfect for this example, and I would bet there were a lot of projects named the same around the time of the movie release.

This project involved every division and department in the organization. Many different new products were bundled together, new customer web interface, new application programming interface (API), new back office release, and network upgrades to most systems. Don't worry about what all this means, but just sit back and imagine all of the silos that come together in this one meeting, each with an inherent agenda to appear as if they are on top of their piece, and with a goal to come out of the meeting without taking any heat. This project moved sweetly along at the PMO group's desired pace. Every meeting flowed through the agenda like a parade of newly minted soldiers smartly stepping across the parade ground in unison, never faltering to regain synchronized walks, and no one glancing to the right or left. We were on pace to set a great record, and doing great things.

Friday afternoon comes on the release weekend and we all come together for one final great moment, and we begin to walk through the final checklist. As we hit a simple web interface the PMO representative calls out on the checklist the network administrator responsible for deploying the code. Quickly the network admin notes in passing with a quiet voice the words that sounded something like, "Hey, you guys did get the new Linux server right?" Really? Friday at 5:00pm on the weekend of the release we ask an offhand question like that? Shouldn't there be a milestone that says purchased hardware, or set up hardware, or provision hardware, or something along those lines? Would we expect that a project wiki, or spreadsheet, or milestone based Gantt chart would be discovered by an interested network admin?

Teams and individuals need the ability to drill down into projects and see a level of consistency, or inconsistency, in order to create for ourselves an expectation for the future of shared engagements. Being able to discover information, or that information is missing, is a key to building cognitive based trust, because cognitive based trust is a decision we make to prescribe trust to an individual, group, department, or company. Maybe the missing server would not have been caught in the dashboard reporting, but at least a connection would have been made regarding the dependent project.

So now we can get back to the discussion about building effective communications and continue with building the plan. As you can now see, I am trying to illustrate a communications plan based around metrics reporting. Communications need to take on many different channels. Remember our communication model early in the discussion about opening up communications? I put it on this page again so that we can talk about channels before we finally get back to building the cognitive based trust communication plan.

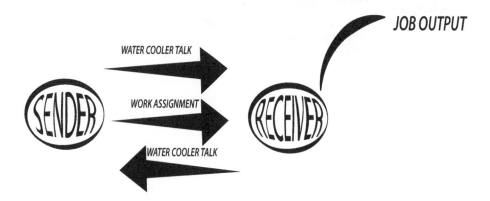

Figure 4.2   Framework of effective communication

It seems that most research into communication channels took place prior to the discovery of the World Wide Web, email, and texting. Unless of course the research is into the question whether we are more likely to crash our vehicle while texting than while watching where we are going. It is, however, a very good metaphor for project communications. If we are not watching where the project is going, we will crash. This goes for our ability to trust those with whom we are working as well.

We need to wrap all available means around our communication strategy, and this includes making all of our information available to discovery. So as we continue to build out the strategy keep in mind where you will want to use wikis and blogs, and web based data driven "dashboards" capable of supporting searches. And then, finally, making sure that we are making available both good and bad messages. We as managers must always remember that trust requires that both the good and the bad are discoverable, and consistently available for it is the consistency, even with a bad message, that builds the ability to prescribe trust.

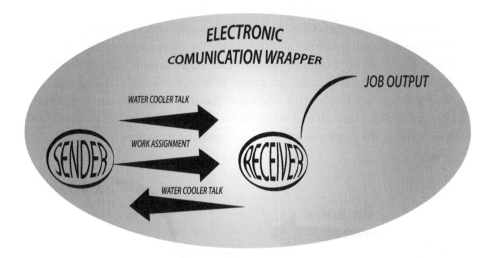

Figure 4.3   Communications need to utilize all available channels

So now back to the build-out of our strategy. Following the standard goals dissemination process, take it to the next level. We use the goals definition process to drive our communications. What do these topics mean to your department, and your projects? It is not just individuals that work remote from the team, but rather entire groups, departments, divisions, and companies.

The next step in building out the strategy has to do with identifying those activities in your organization that may have an impact on the overall goals (see Figure 4.4). This may be as simple as listing the activities that are currently taking place, or brainstorming the activities that will take place later in the year. Always start with a place holder for the tactical level goals making sure they meet the requirement of being Specific-Measurable-Attainable-Realistic-Time (SMART) boxed. Keeping this principle in mind keeps people involved and motivated to achieving the goals that are set.

Tracks are those activities that may be following a series of events designed to achieve a specific outcome. This may be a series of projects toward building a knowledge management system that provides for the capture and categorization of department experiences designed to mentor new employees, guide problem solving work, or share project based knowledge with remote employees. This process is great for providing a type of peer training or mentoring across the great divide that can develop when team members are working in the virtual environment.

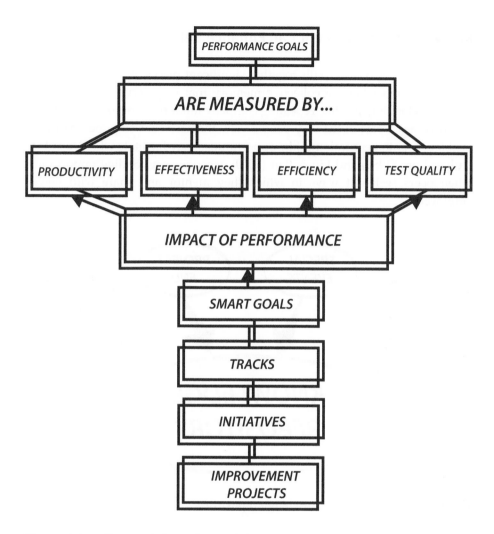

Figure 4.4 **Determining what activities may impact the communication strategy**

The people questions can be fairly easy if we followed some of the suggestions in the personality based trust sections. Now we simply need to tie some of that information into the search capability of a dashboard or project measures and metrics, or better yet, a knowledge repository. Helping people remain concise and relevant in their information discovery is very important, and tying this back to the project will help in keeping with the targeted information flow.

Knowledge Management Can Enhance Belonging

A knowledge repository pulls together significant pieces of an organization's experience, and creates a collective memory that may serve to enhance the sense of belonging. It is like repeating company folklore at corporate functions, and becomes a part of the way people view themselves. Using the process to facilitate peer to peer training-like opportunities may facilitate what Dossett & Hulvershorn (1983) and Liu & Batt (2010) describe as a bonding opportunity when peers seek out ways in which they mentor and share with newer employees.

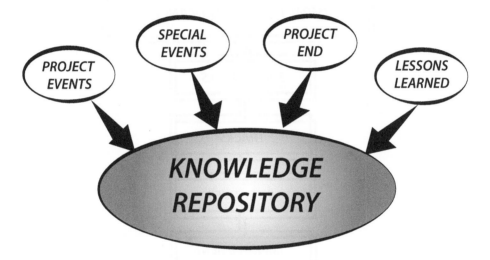

Figure 4.5   Building a knowledge repository

The repository becomes a place where experienced and knowledgeable employees reach out to share their wealth of knowledge with new employees, or up and coming new leaders in the organization. As the repository is built, using the experienced employees to vet and organize the knowledge as subject matter experts (SME) allows for refinement of the knowledge transfer process, and aids in identification of those pieces of information that may relate to greater categories providing detailed captures of critical corporate assets. Assets that otherwise may be lost in the day to day activities, or that simply don't transfer to remote team members.

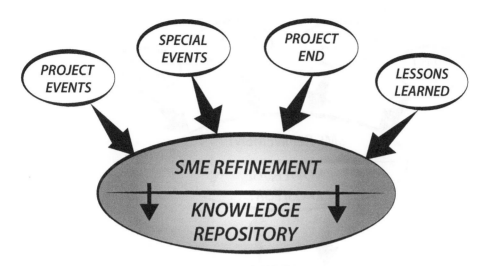

Figure 4.6   Using Subject Matter Experts to refine the knowledge repository

A knowledge repository then becomes a primary feeder of corporate knowledge into the communication stream as a means of information discovery (see Figure 4.7). Employees and team mates can mine the repository for "corporate lore," as it may be called, regarding the past experiences and knowledge of team mates and the corporate giants that have come before them in paving the way to the words "how we do things around here" (see Figure 4.8).

Now it is time to keep building out our model. Tracking the special initiatives that the organization is undertaking, special projects, and improvement efforts, as they maintain a line of sight to the goals helps in communicating the efforts of the organization. Tie together the special efforts with the SMART goals, tracks, and initiatives into a common framework, or repository, and measure their ability to affect the goals.

As we noted in the beginning, cognitive based trust builds trust in the three categories of unit grouping, reputation, and stereotyping. The unit grouping piece is based on the process in which team members work to attain common goals. We have talked at some length regarding the tracking of goals. This is to ensure we build the ability to identify those common goals in such a way that those working apart from the main group, which by the way may be most of the team, are able to watch and track the progress toward those goals.

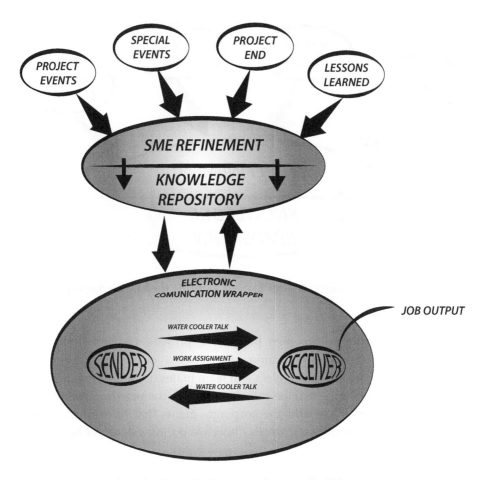

Figure 4.7   Using the knowledge repository to build trust

Of course, in most cases, team goals are very tactical, however knowing how they relate to strategic goals, and how they affect the overall team can have a great impact on team identity. When the tracking is drilled down to the next level, it allows team members to see their direct line of sight to the overall team, and in many cases, how their work directly affects the bottom line, organizational goals, and bonus attainment.

At this point in the reporting process we have now developed a full picture that can guide, and ease, the ability to discover relevant information for all team members. The information discovery process is guided and can easily be navigated throughout the entire organization allowing for team members, when questions arise, to develop an effective picture of their team, individual

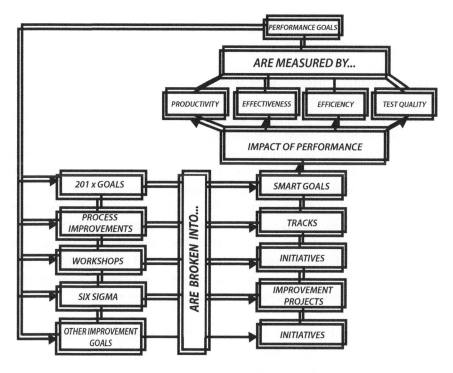

Figure 4.8   Bringing together special efforts in common reporting repository

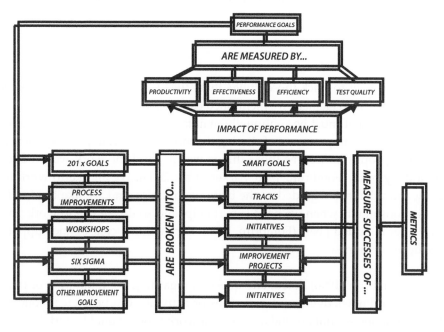

Figure 4.9   Tying in metrics with the common reporting repository

members, and organizational peers. This is often a very real problem for most organizations (see Figure 4.9).

I know recently a problem arose at work with one of my peers from another organization. This individual wrote a rather scathing remark to my boss about a project with which my group was assisting my peer. I decided before I talked with this person that I wanted to figure out exactly who this was, and what their role was, but there was no information available beyond his name and a note of to whom he reported. I had no idea what this person's responsibility was, what his goals are, or why he might be so agitated, and no way to prepare to talk with him on a level that would establish a degree of trust before we began the conversation. If we had available to us a combination of the organizational structure, a personalized wiki page, or perhaps the corporately developed social networking site with a bit of personal information, and the project level goals tracking, we would have been very well prepared to figure out the problem and come together with this person in a problem solving mode.

Instead the problem was approached from a perspective of mistrust and speculation, and the phone call went immediately from agitated to confrontational. It took most of the hour long call to move from confrontation to information discovery, and then problem solving. This is a problem that can be avoided with the proper information discovery structure we built here.

Reputation

We always want to track those things that affect how other groups or people view us. This is an element we mentioned earlier about reputation. Reputation is something that must be continually maintained. A reputation is built upon the word of mouth processes of sharing the information from past experiences of many people, as well as those things we say, and the work we produce, on a daily basis (Fombrun, 1996; Sarker et al., 2003). We all like to work for companies that are regarded as the better companies, and we like to hire the better consultants when we need expert help, and so do we like to work with the better regarded peers (Fombrun, 1996). Reputation increases and supports the decision to trust.

As Fombrun (1996) noted, any question regarding the reputation of those working in knowledge based industries or activities can be a great concern. Our reputation is built around a consistent work product quality and delivery,

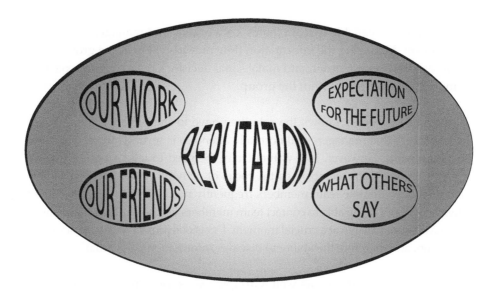

Figure 4.10   Building a reputation includes work, relationships,
expectations for the future, and feedback

and the expectations for future engagements at a given quality. Reputation is built upon the interactions we have with one another, along with those experiences that others report. We then create for ourselves an expectation of what a future interaction may be like with this person. This expectation we now have becomes the reputation of that person.

When we talk about a reputation, what we are expressing is what may be generally accepted as the expectation for the person's behavior, work ethic, or work output, for example. This information is then combined with any misinformation encouraged by gaps in the information stream. Maintaining a continuous stream of available, relevant, and correct information is a sensible strategy for avoiding reputation risk.

Earlier we described connectedness as the sense of family within the group. As we evolve the discussion in relation to a decision that is made to trust or not trust we understand connectedness as related to the validation within the group of one's reputation as acceptable and positive (Polzer, Milton & Swann, 2002). The degree to which we receive respectful treatment from the group, however, may be related to how well we value our reputation within the group (de Cremer & Tyler, 2005). Thus, as noted previously, the need to ensure that the information available to team mates, co-workers, and managers is positive and relevant. Managing the information flow may also help as the group seeks to sort out the pecking order, and make some decisions regarding the role that an individual may play within the group.

Everyone generally knows the role they have in the project group. One of the first things that managers will normally attend to is sorting out the skills and job assignments. However, in a virtual team, the need to take a leadership role can shift among team members as the focus of activity moves between remote team members. Remote team members frequently need to make decisions about activities without being able to contact team members whose working hours mean they may be unavailable at a crucial moment. Decisions impact the team member's reputation, and therefore the ability to discover information is essential.

This is true for groups as well. Groups, teams, or departments have the same need to take charge at different times, depending on the stage or lifecycle gate in play; making decisions outside of the larger group. One of the keys to making sure the reputation is maintained is to ensure effective management of the information. Information management is an essential element of the communication plan. Identifying the stakeholders involved in a project or decision making role helps in determining to whom we need to communicate, and what information should be communicated to each stakeholder. The communication may be as simple as sending the link to your information dashboard, or ensuring specific metrics are provided on a regular basis.

Stakeholder Analysis Process for Planning Communications

The stakeholder analysis process is pretty simple and widely available. Brainstorm and record a list of those individuals and groups that will affect, or be affected by the project, initiative, or activity. This may be those working within the process, those that feed the process, or consumers of the process output. This may also need to include those that are simple observers of the process. Once this is accomplished begin to map their specific interests. Output of the analysis can be posted to a spreadsheet for ease of organizing the thoughts.

Managing your personal presence in corporate communications, web sites, and other tools is also an essential tool. Decide what interest the stakeholder has and what changes the project may place upon their interests, and how important their interest is in the process. Once you have done that, decide what role the stakeholder needs to play in the project, or if their reaction may be positive or negative. This preparation will help you decide the type and style of communications, the frequency of the communication, and the content. Taking these steps will help the team understand how to manage the information flow better, and therefore help the team, group, or department manage their reputation. Remember, your aim is to avoid surprises, misinterpretation, or miscommunication.

Individuals can use this simple process in determining new ways to manage their own reputation. Stakeholder analysis works in both directions. It is a way of determining what information someone may seek about you as well as the information you need about them.

Stereotyping

When we employ a stereotype we are accepting impressions about others, or groups of others, based on appearances or other interactions, and form assumptions about traits that may co-occur in groups of others (Baldwin, 1992). The impression that someone may have about us will affect their decision to trust us, or not trust us. Because of the influence of stereotyping, the impression that we may have on others must be actively maintained and managed and tailored to affect the image of a "consistent," "rational," individual that may fulfill the "role requirements" (Schlenker, 1975).

All of us will make assumptions about the people we work with and we will use stereotypes, or previously developed mental images, in doing so. There are times when a stereotype may have a degree of accuracy and times when it can be entirely flawed. Make sure that the impression and information you make available about your work, your communications, or other information, is as consistent, accurate, and informative as possible.

We are all very familiar with social stereotypes. The mental imagery that arises when talking about work groups, organizations, or company brands are a little less obvious. Organizations and particularly brands work hard to establish simple, programmed responses amongst customers; effectively they encourage stereotyping. Consider the corporate technology store Best Buy Company, Inc. trading as BBY on the New York Stock Exchange with their trademarked mobile assistance role named the Geek Squad. The services provided by a small group of technology geeks working for Best Buy is now strongly branded, with services preceded by a reputation that brings about a common visual and skills expectation. They like it when their customers think that a technology obsessed young person with nothing on their mind but enjoying a few moments with their new computer to optimize it beyond their wildest dreams will soon arrive on their doorsteps.

But what happens when the expectations rise above the capability of the team? Perhaps the hype raises the expectations that the team may have for themselves straining the ability of the team to live up to their own expectations. As Cocchiara and Quick (2004) point out, even commonly held strong positive expectations can have a negative effect on a group or department. Overwork, burnout, and team strife are possible outcomes when a team drives too hard to live up to the expectations of others.

Negative expectations of a group's performance can lead to the same outcome as overly positive preconceptions. Increased efforts to refute low expectations may drive a team too hard causing breakdowns in capability, causing mistakes, just as when driven by high expectations of performance. The stress incurred in the attempts to refute the stereotype can then lead to sustained reductions in performance over time (Cocchiara & Quick, 2004).

These stereotypes can be aggravated when working in a global virtual environment. Expectations that a national group will have greater engineering skills, or greater exposure to computer technologies, can add stress to the group experience as those within the group involved seek to live up to the unreasonable expectations of others. Worse still can be the expectation that a national or ethnic group lack the traits or behaviors desired for the project.

Somewhat confounding to the discussion is the notion that these preconceived ideas may be attached, not to an ethnic group or nationality, but to a corporation due to past business experience with the company or to market performance. I have seen this happen more than once and it is damaging to the organizational relationships, to the people, and to careers, as well as the corporate reputation. On one occasion I worked with two major financial institutions that came together to form a new joint venture, and on another, on the merger of two major financial institutions during an acquisition. The two situations were different but the problems were similar and equally damaging.

In the case of the merger, the acquiring team, as may often be the case, began with low expectations for their new team mate's skill, and in response people worked long into the night to establish themselves as strong performers. The zeal to overwhelm their new peers led to exhaustion and mistakes. With the joint venture the competition to establish dominance for one team over the other caused many delays and mistakes as the project progressed. Delays in the schedule were often caused by rework as the teams pushed each other hard as they competed rather than cooperated.

A Corporate Stereotype That Caused Elevated Employee Stresses and Delays

I have been involved in a couple of mergers at this point in my career, as have most of us. The stress of having to prove oneself to the new team and the new boss can drive a person to stretch their capabilities and knowledge. Change is stressful even if the change is considered to be for the best for all involved. I had joined the company only a few months earlier knowing there was a possibility that a merger was on the horizon. It is always a good idea to do a little research before changing jobs, so I had scouted around to see if change was on the way, and had discovered several previous attempts at building a merger offer and was prepared for change. I was also pretty confident that I was going to be joining the underdog if a merger occurred.

In a merger or acquisitions situation, often members of one of the organizations may have preconceived ideas about the capabilities of the other. This may to the result of previous joint experiences or perhaps simply because of lack luster market performance. Remember that we talked earlier about how experiences create expectations of comparable performance in the future, and how groups of people may be tainted by those expectations. In this case the entire organization had been tainted by regulatory problems from the past, perhaps caused by shady dealings, as well as technology problems and failures.

We were acquired by one of the giants in the industry, and this company used only proprietary technology, a completely new information technology (IT) architecture, and a completely different development process from what our team was currently accustomed. When these teams came together, we, the new guys, had to learn new technology and new methods of working together, seemingly overnight, and at the same time participate in a complete corporate rewrite of primary corporate technologies on a very tight schedule. Our new bosses and team mates had very low expectations that we would be able to keep up, and yet we had to keep up to keep our jobs. Big stressors here. Developers worked 24 hours around the clock for days and weeks in an attempt to overcompensate for the low expectations. I would see emails sent from team members that were still working at 2am and 3am. After weeks of this behavior they would finally collapse, exhausted and frustrated as they tried to simultaneously learn the new technology and meet very tight schedules they set for themselves.

Mistakes, as one would expect to occur when teams work themselves to exhaustion, began to take place. Communication between team members dropped, and collaboration between the teams fell apart. The drop in capability, a result of overexertion as the acquired team attempted to impress the new owner, and resulting increase in stress, damaged the project team performance. In the end, while the project came in on time, the product quality suffered and the project launch was riddled with application bugs.

How Do Stereotypes Affect Cognitive Trust?

As we talk to and work with others, these formal and informal interactions are collected and cataloged as good or bad. These experiences are turned into expectations for future interactions that we hold for different work groups or groups of people (Sarker et al., 2003). We all use this information to make decisions about how we expect further interactions to play out. The decisions we gather about other people or groups of people form our belief system, and help us to decide whether a person or group is able to be trusted, or trustworthy.

Combating Stereotypes

The real problem is that preconceived expectations, whether they are talked about openly or not, can affect how well teams work together. Even social encounters with virtual team members affect our perceptions of trustworthiness. How can we deal with the problem of stereotypes and the way they can degrade trust in the work place? Trust is affected when performance is inconsistent and information is difficult to find. As you remember, the process of viewing ourselves as a member of the team is affected by how we identify with the behaviors and traits of the team, and in this case, the process of self-affirmation affects how we identify with the stereotypes (Martens, Johns, Greenberg & Schimel, 2006).

So what is a manager then to do with this knowledge? Remembering that positive stereotypes encourage trust and enable better collaborative performance, a manager must control the information available to ensure a healthy stereotype is developed. Additionally, each person, whenever there is an opportunity for interaction with other team members or groups, must take responsibility to ensure they are always presenting themselves in a manner that develops a positive viewpoint.

This can include taking any opportunity to ensure a positive story is made public in measurable ways such as weekly performance reports. Find ways to display and reinforce positive expectations. This may mean establishing performance indicators that reinforce the good performance and a consistently present a good picture of the individual and the department or group.

Group stereotypes may have a similar effect on an individual's performance in the work place to ethnic or racial stereotypes. Groups such as quality assurance (QA) or IT test, according to Weyuker et al. (2000), often suffer a negative stereotype. Groups with a negative stereotype may, for example, be perceived as less technically capable and with a poorer grasp of the technologies than other project team members. By some authors, they have been at times, described as working in opposition to the development team by seeking ways to sabotage product deployment, rather than working together to get the product out the door.

There are a couple strategies that have been shown to work in combating the effect of negative preconceptions. This is even more important when the team is working in remote locations, distant from the core team, and therefore may be less able to address effectively the negative stereotype on their own. Information provided in the organizational metrics portal should allow individuals to follow up on their own performance, and support managers in presenting solid evidence of solid performances. The point of this discussion though is the effect that this strategy may have on trust. So how does this all tie together to affect cognitive based trust?

We briefly noted at the beginning of this discussion that cognitive based trust is comprised of inputs being of unit grouping, reputation categorization, and stereotyping. All three of the elements require the gathering of information in order to make a decision regarding the prescribing of trust to another. Stereotyping accounts for the activity of taking casual encounters with others and prescribing a positive, or negative, stereotype.

Figure 4.11   Cognitive based trust model with data inputs

Putting all of this information together provides enough detail for a person to make a decision as to whether they should or should not prescribe trust upon another person or group. How a person, department, or company manages this information can determine the effectiveness of the project team or group.

5

Building Institutional Based Trust

Everyone has strong opinions of some sort as to the trustworthiness of a company. As I begin writing this chapter I keep looking up at the TV to see protesters carrying signs for Occupy Wall Street. Whether you agree with the protesters or not, the message is clear. Many people simply don't trust companies. Institutional based trust is determined by the perception of fairness and equity in the way a business, group, or department, and their representatives, transact or implement corporate policy.

We need to be aware that every action taken by a manager, and every email, text, or tweet that a manager may send, and every discussion that a manager has with an employee or co-worker, will affect the perception of corporate trustworthiness. It is a common understanding that the norms and rules of an organization has a direct effect on the way people do business, and therefore, the business that we conduct reflects the fairness of the institution. Thus, the way the Occupy Wall Street protesters may be perceived to see the world of big business, since big business in this case is portrayed to have unfairly profited in the current economic collapse, the system that allowed them to profit must be inherently unfair.

This same process of conviction by perception, albeit on a much smaller scale, may play out every day in every work group, project team, and organization. When a co-worker or team member experiences an event or corporate program, that experience can carry with it an expectation for the next event or corporate program. The events of one's day are never experienced in isolation as a unique occurrence, but rather as a potential for future events and experiences.

If you remember way back in the beginning of this book we talked about my experiences as a young, fresh faced boy covered in coal dust and sweat. Every day I would arrive at work early to make sure I was ready to hit the hills of coal piled high waiting for me to scoop them up and place them back in their proper place for a morning start, and every day I would pass people on the road that I knew were still supposed to be working. They were smiling as they rode past on their motorcycles, or fast cars, knowing each day that life was good. No one was going to mess with their routine, and yet, I knew that the operations manager was going to be there waiting to see that I had arrived on time.

Now, as you remember in my earlier story, I too would escape as soon as I could every day, but of course in my adolescent brain that was irrelevant. The important part at that time was my arrival, and it didn't seem fair, and I'll tell you why. It seems contrary when we spell it out, but it makes perfect sense in my head. Events occur in isolation, but are experienced in light of the coming future expectation for the next event. I saw them leaving early, and although I knew that I too was probably going to leave early, passing them on my way in meant that I had to be on time this morning, but they got to break the schedule.

So now for the kicker. According to Sarker et al., the belief that a company or institution requires their people to follow the established rules allows the employees and project team members to trust the other folks that work within that system (2003). This trust may be extended to people that we might have never met face to face, and may have no expectation that we will ever meet face to face (Sarker et al., 2003). This expectation of trustworthiness can be prescribed to an entire institution, company, work group, or project team when the rules are clearly identified and known by all, and then enforced. Now think about the protesters. Their perception of trust was violated when they believed they had discovered that the rules don't apply to everyone.

While average people are losing their jobs, and maybe their homes and in some cases their families, and they watch news reports of the economy falling apart due in part to banking system "irregularities," and seeing reports that appear to show that government rules for failure don't seem to fairly apply to all, is it really a wonder that the protests began. The protests first began with the Tea Party protesting big government and the bailouts. Now we see the Wall Street protesters protesting big bank bailouts. This news reflects a massive failure in trust of our government and legal institutions.

So as we extend this discussion to our work place, what does this mean as a manager responsible for implementing and enforcing the rules of engagement? As we start this discussion, every time we open our mouths or send a message we are sending a message about the trustworthiness of the entity for which we work. Even the smallest interaction with anyone else carries a message regarding the trustworthiness of the system.

Here is a point that we all tend to settle somewhere in the memory banks. Fear of consequences too can build trust. Every child knows this for certain. The fear of punishment builds trust as well, for every child knows that if they break the rules they may trust that there will be consequences, and therefore they can trust that the system is fair for everyone. But what about the child we all knew as being fearless? They never worried about being caught. Now that some of us are parents, we may now understand why they didn't worry about being caught. They did not live under the same cause and effect paradigm as the rest of us. Oh, by the way, these are the kids that we watch as adults and try to keep our own kids from playing with them too often, or going to their homes, because the lack of a cause and effect relationship regarding behavior and punishment for breaking rules can create an environment where bad things can happen.

These are the same consequences we may forget are very important for building trust in the work place. Not only do we need to enforce the rules fairly to be sure our employees see the company as being trustworthy, but we also need to carry through with the threat of punishment when the rules are broken in order that our employees see that the rules are real and can be trusted.

All of this rolled up creates an environment where institutional based trust can take shape. See Figure 5.1. If a piece is missing, then trust is shaky, and that big gaping gap that we talked about in the past chapters will open. Folks begin filling the gap as they see best, and bad things may begin to happen.

Good things can happen, though, when we get this part right. Remember that virtual team members don't have the experience of face to face interaction with other team members, and therefore are working mostly from the perspective of the written word. They don't see the grey areas the same as those working face to face. They don't see the nuance of facial expressions and body posture, and may therefore merely take the rules at face value. They are often missing the stories and corporate lore that is shared in the meeting room and around the coffee pot.

The rules are inked into the pages of the online handbook. Rules are reinforced when human resources' emails are received when someone decides

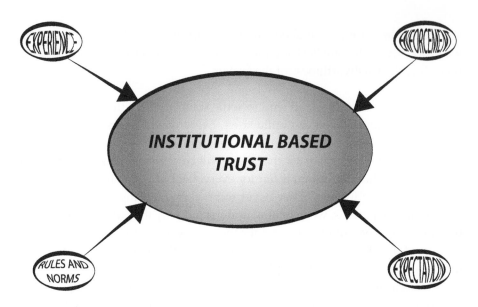

Figure 5.1   Institutional based trust model with inputs

to restate or update the rulebook. Rules are further reinforced when human resources sends out the promotion notices, or not. Notices sent to team members about holidays, corporate parties, and team breakfasts tell their own story about fairness and equity in the system. All of these innocent and sometimes automatic communications and events tell a story of their own for virtual team members. Team members that work in a virtual setting, apart from the rest of the team, can strengthen or weaken their sense of membership with the team, and learn to trust or mistrust, when they participate in the established norms and practices of the organization (Sarker et al., 2003).

What Does Institutional Based Trust Mean for Projects?

The basic rules of engagement for projects are the same whether we are talking about information technology (IT) or mechanical engineering projects, and everything in between. As a company or department we establish a set of rules called the lifecycle, display these rules for all to see and live by, and then go on our merry way expecting great things to happen. Every company has a set of rules on how things get done. The product lifecycle, or service delivery lifecycle, as it may be, establishes practices that the team is expected to follow. Requirements are identified, whether on a napkin, in a formal online tool or spreadsheet, or in a notebook, but we begin with identifying what we want to accomplish. Requirements are affirmed, and teams set in place to build or deliver based on the requirements. Stage gates, either formal or informal, are a part of every lifecycle. Someone reads the requirements and decides if this is something that can be delivered, a level of quality is determined, and schedules and budgets are identified to make things happen. Somewhere in the process these project elements are reviewed and agreed to by all parties, and work begins.

Basic lifecycle elements remain the same, and tend to be pretty well globally accepted. As the project progresses the schedules slip a bit, but not always too far. Budgets get a bit stressed and at times maybe resources get renegotiated as the pressures build, and then we begin to find out how much of a commitment we have really made to the rules. Sometimes the company norms allow for the rules to bend. Maybe these norms are OK, and generally reflect how work gets done. Maybe these bendy rules are the norm and everyone can pretty well trust that this is acceptable, and everyone gets pretty much the same treatment.

A problem can occur, however, when we establish a set of rules on how work gets done, and then have a tendency to bend and stretch, and adjust the rules based on unpublished pecking orders and hierarchies. This may include unwritten rules such as software development being an art, and therefore cannot be held to strict rules regarding timeliness of deliverables or product specifications. Maybe the development team, although the rules say they must document the specifications, can't really take all the time it would require to write everything down, so … sometimes … maybe …, OK this time only documentation can slip if needed to meet the schedule. The team downstream now is expected to work with a little less information. Hopefully they can be counted on to take up the slack and tighten up the schedule to make up for the missed deadline. Often this is the product test team, and everyone knows they have a tendency to throw in obscure and highly improbable tests, so the test plan and strategy had better be on time to allow a thorough review for these unreasonable attempts at search and seizure of developments hard work.

Sound a little cynical? Well, that depends on which side of the upstream–downstream product lifecycle your career has grown. Fairness and equitable application of the corporate rules is very experience oriented. Managers must always remember that, like the child we talked about earlier, everyone is watching to see what mom and dad are going to do when the rules are "bent" a little. Everyone is watching to see what happens to the team that tends to do the bending. When the promotion announcements go out, are the benders on the list, or are the teams downstream that always work weekends and holidays to make up time making the list? Are there names on the list of those that other team members see passing them on the road as they arrive to work early for the next shift?

Problems on projects will always lead to hard decisions. As managers, we really need to be sure that these hard decisions are consistent, and support the needs of the entire project team. We also need to be sure they are always made consistent with the goals we talked about in the last chapter. Our stated goals should be at the forefront of every decision. If a decision does not support the attainment of our goals, then perhaps we need to take another look. If we publish a goal that says quality must be improved, are we then making decisions that reinforce the quality rules?

Let's consider a situation that probably happens more than we would like to admit. Once upon a time I worked for a pretty big player in the financial world. We liked to think we were good at what we did, and so every year we would ask our customers how much they loved us. Pretty simply put, our annual customer survey told us once again, that they think our product stinks, but we are the only game in town, so they play, and hate us for it. We finally began to get their point when one year it was decided that we should go public with a stock offering. Now this changed things a bit, because if we were going public we would have to have customers that really loved us and our products. Not the fictitious love that we told ourselves they had deep down inside where it really counts. So, to rectify the situation we established goals to improve the quality of our products and services.

We set off to do great things, started up a bunch of very optimistic and demanding projects, and went forth as a team to make great changes to the world around us. Software developers were let loose upon the thoughts and ideas set before us. We wrote the requirements down in large verbose documents pontificating on the subject of financial strategies of the future, and …

WAIT. OK, strategies were developed, and … WAIT. The quality team began to stomp their feet and scream, "Wait for us!" Software developers don't start coding before requirements are written and strategies are developed. The quality team frantically began reviewing requirements and strategizing their testing, and commenting on the testability of the requirements as the project continued. Testing started as the project moved along, and big problems were found in some critical areas. Development pushed back a bit, but things moved on through the schedule, and more problems were found. Now the development team started to get a little heated since the schedule had moved pretty far down the line to be finding such big and critical defects. How can they meet the project schedules if critical defects aren't found early?

The schedule had to slip now because quality should have found these defects much earlier in the process. WAIT! Quality tried to say real early that testing would be difficult since many requirements weren't defined clearly, and testability was a question. OK, any way testing would compress and take up the schedule slack.

Fast forward and here we are now on the release weekend. Everyone is proud, maybe quality slipped a bit, but we made it, and final verifications are being made as development, quality, and ops are all gathered together to see the final product prepared for launch. But WAIT! What was that? That's not what the product is supposed to do! Really? Yes, the proverbial rope swing swings straight into the tree trunk. Way back when, when the requirements were written after development started developing, and the requirements review was dropped to keep the schedule we had all made independent and different assumptions about what requirements really meant. Testing passed because we tested what we thought should happen, and development built something close, but not really the same.

What Does All This Mean to Us as Managers?

We must align our goals, our rules, and our actions in order to establish an environment that can build and nourish institutional based trust. We have to align our actions, goals, and measures of outcomes so that a consistent and fair practice emerges, and our employees and team members feel and see the consistency.

This is not always as easy as it sounds. We all know that old saying, "One awe sh_t erases ten Atta-boys." It is very true that giving in to temptation to make the schedule at the cost of stated goals or broken rules will erase many of the gains made in establishing institutional trust. So what can we do?

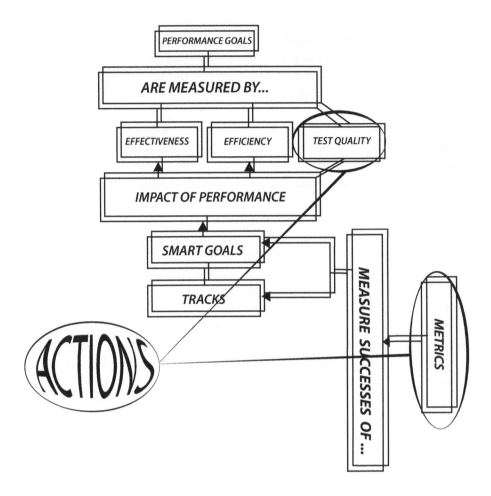

Figure 5.2   Aligning actions with goals and measures supports institutional based trust

Quality Assurance as a Way to Strengthen Project Processes

We need to put in place processes that track and manage the rules we have stated. This can be done through the use of formal or informal quality assurance (QA) practices and reporting. QA as a process that identifies and then assures compliance to the activities, artifacts, and actions needed to fulfill process commitments is one means by which we can be sure that what we say we are going to do, we actually do.

This process should also be accompanied by reporting that makes compliance, or a lack of compliance, public knowledge. Remember that the ability for people to discover information is extremely important in the support of trust relationships. It does not do us any good to do the tracking and management of the processes if no one is able to discover the extent to which we actually comply. We want everyone to know how well we are doing with compliance.

Good and bad compliance, when consistent, can still build trust. Good compliance, though, builds strong bonds. This can be done in several different ways. Process artifacts, or templates that capture project milestone documents like requirements, schedules, test strategies, and so on, can be created and maintained using forms. Forms can then feed a database that can be used to report on the degree to which the process is fulfilled.

Simple reports can then be auto-generated to tell everyone how well a project is complying with the stated lifecycle. The reports can be held in a standard repository that everyone can access for all the projects and get immediate feedback as to how the process is functioning. This type of openness and ease of discovery can provide effective and immediate feedback to all employees as to the equitable implementation of company policies. This provides quantifiable evidence that we are all equal and share a common set of practices no matter where we sit. Trust is supported by common norms and practices, and while very basic, the practices that define how work gets done is very important.

Another easy to implement practice is the use of online survey instruments and web sites that can be used to ask project participants about milestones and artifacts. Simply asking the question and then following up on the non-compliances can go a long way. Just the fact that someone cares to ask, can help. Further following up reinforces that someone cares about the outcome, and reinforces that everyone is part of the same team.

Free web based survey sites offer access to online survey builds and delivery mechanisms that can be used to customize a quality program. The information can be packaged and reported directly from the web site, and the information displayed for everyone with very little effort and resources. QA is designed to address the question as to whether the lifecycle is effectively implemented, and that members working within the lifecycle may be successful in producing a product that meets the demands of users as defined in the requirements

(Wise, 2011b). A primary goal in this strategy is to access the lifecycle in such a way as to avoid intruding upon the daily activities of the project team.

Remember the goal here is to build a clear understanding that the policies and procedures of the organization are equally and fairly applied to all participants. By making public how the lifecycle is managed, and the degree to which participants in the lifecycle are expected to comply with the policies and procedures, all participants are then able to gather the information to confirm an even playing field. But what happens if our publicly displayed information shows that the process is broken, and no one is following the process?

Trust, we have to remember, can grow even if the trust is based on the knowledge that we can count on everyone cheating the system equally. A better plan however is to build predictability into the processes, and thus the great need for a solid QA program. Predictability is very important in building institutional based trust, and may be described as the degree to which the leadership is willing and able to reward the behaviors necessary to meet the defined goals, and the degree to which the norms and values of the organization are reinforced (Gillespie & Mann, 2004; Sarker et al., 2003).

You see that defined goals piece returning over and over to the conversation. Defining, measuring progress, and making the progress toward the defined goals very public is a lynch pin in building trust of all kinds. It is the beacon that guides the entire organization, and becomes the playbook by which everyone knows how to act and what to do.

Maintaining order is a big deal in building predictability, and so, as orderliness strengthens predictability, then the need to reinforce the desired behaviors may not always be a pleasant experience, thus correcting the non-compliance does build trust (Sarker et al., 2003). In other words, trust, in part, is the ability to expect that the future interactions will be the same as the experience we had today. It is the expectation that behaviors will be consistent and desired behaviors reinforced.

When working in a virtual setting the need for feedback and assurance on the state of the process, or lifecycle, becomes even greater. People trust, according to Pavlou (2002), because assurances are provided that systems are in place to ensure an even playing field. If we want greater assurances due to a transitional state, or times of greater uncertainty, then perhaps calling in outside experts to certify the state of the process may be warranted (Pavlou, 2002).

There are plenty of companies scattered around the world capable of performing a process certification. Most consulting firms have developed their own process maturity model and are very well versed in applying them to this situation. If you want to go with more recognizable standards, then CMMi from Carnegie Mellon University, or perhaps international standards such as ISO may be more applicable. In my case, we simply adopted our own model and performed our own assessment to provide the baseline, and began building and strengthening trust from there.

Institutional Trust on a More Personal Basis

Never forget the need for predictability and the impact consistent behaviors can have on how employees, co-workers, and team members perceive the fairness of company decision making. In order to provide assurance regarding decision making, when it comes to individuals processes must be put in place that, once again, make this information available to discovery. The problem is how to accomplish this when personal privacy is a concern.

Human resource policies normally require that personal information not be shared, but the processes that guide human resource policy should be open and apparent to everyone involved. Like the discussion about QA, the need to make this predictable and consistent is essential in order that everyone can see the level playing field. In this case, the level playing field is built upon, here it is again, the goals. Yes, human resource policy must also map back to well defined goals. Goal setting, while it begins way back at the beginning of our discussion, has to map all the way to the end in our discussion of institutional based trust. In this process we build the consistency into everything we do, and calm can come over our little world.

Mapping our institutional goals down to individualized goal setting levels the playing field for everyone, and provides a process by which personnel decisions of promotion, compensation, and discipline can have a common framework. If we use this process correctly, when employees sit down with the manager there are no surprises and everyone has the same expectation before the process starts. Managers no longer have to struggle with deciding how to deliver a difficult message, and employees know ahead of time if that difficult message is coming. Problems occur only when the message doesn't match the expectation.

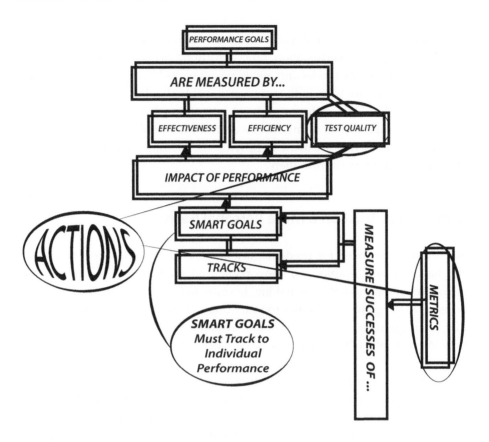

Figure 5.3   Aligning actions with goals and individual performance supports institutional based trust

With this in mind we can take the defined organizational goals, build in performance improvement efforts and tactical plans, and create specific attainable goals for each individual. These items need to have specific activities with planned outcomes and dates for expected achievement. This information should be documented in each individual's yearly planning document and tracked on a regular basis.

The key to making this work, however, is to remove the subjectivity in the year end review. Human resources, by establishing the process, and then tracking and managing the process can eliminate the common failure in the annual reviews. People don't like to deliver bad news. Annual reviews are a process just like any other process, and a direct and immediate reflection on the trustworthiness of our institution, thus a key factor in establishing institutional based trust. So, have a QA function that provides assurance that everyone is once again standing on that level playing field. Human Resource is in an ideal position to be that QA viewpoint into the annual planning process.

Every annual planning document should be reviewed for certain elements. Is the goal Specific-Measurable-Attainable-Realistic-Time (SMART)-boxed? If the answer is yes, then require the individuals, not the managers to provide evidence that the goal is met. People love the process if it is perceived to support the fairness that everyone desires.

Using Performance Based Evidential Evaluations

Time for another story. Once upon a time I worked as a line manager in a nuclear power station, and along with a few co-workers, set about trying to fix a process that was completely stifling our ability to get work done. This process required hourly union employees and management folks to work hand in hand to accomplish some complicated tasks, and it caused the lines between the union and management jobs to blur a bit when accomplishing the jobs.

The blurring of the work assignments was a big point of contention, and expectedly so, for there is no real value in establishing union agreements if we are going to cross the boundaries and confound the agreement in the process. As a part of solving the process problems we decided to bring all of the participants into one department to help everyone get together in deciding how work should get done, and aid each other in task sharing. When we got the group together we discovered that our annual review process simply didn't support working together, because everyone was being graded on tasks that required unique and separate activity. On top of the simple problem of redefining the tasks was the issue of trust. The union folks had to trust that, while maybe we were blurring the lines of responsibility, we would not take advantage of people by grading them above their pay-grade, or raise the bar without negotiation. We decided to try something completely new.

People had to grade themselves, and then provide evidence of the grade in order to negotiate any differences between their grade and the supervisor's grade. At the beginning of the year I established their personal goals, which due to their position in the union, was predefined. For each of the goals I also set up a matrix of behaviors that predetermined the grade level that would be given for each behavior. In this particular case the union members were part of a writing team, so their jobs were keyboarding and formatting of technical documents. I, as the supervisor, assigned a grade and was required to provide evidence of the grade that I had chosen. Whenever a particular behavior was contested, the employee was required to also provide evidence that supported their point of view. If the evidence was such that I was compelled to adjust my grade, then we would negotiate a grade that made us both comfortable.

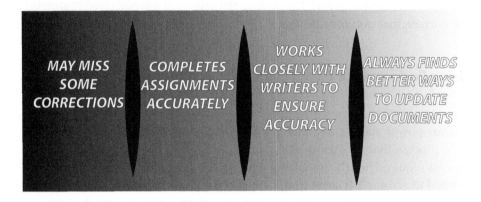

Figure 5.4   Example of a behavioral grading scale

When we were done with the grade assignments we were all able to support, and live happily with the outcomes. I did not have to give someone bad news that they were unable to support, and they, when the grade was adjusted were able to proudly say that their grade was just. No one complained, and no one felt as if they had to justify why their grade was better or worse than anyone else. Now we all know that people are not supposed to share their final grades, but of course everyone does. Everyone shares their grade that is unless they feel the process is fair. When there is a feeling that everyone is being treated the same, the need to compare notes simply melts away.

Just like eating at your favorite restaurant. When you trust that the hygiene issues are well addressed, then you don't feel a need to check out the kitchen before you sit down to eat. Job grades work the same way. I had human resources check out the process to be sure it was handled fairly, and that they could support our new process. The quality check is very important because it provides that level of assurance needed to build trust where transparency simply cannot be expected. Another great suggestion is to avoid the use of arbitrary tools such as the bell-curve.

The Bell-Curve Problem

Many companies like to use a bell-curve to force fit employee reviews. There is only one good explanation that I have ever heard for force fitting employee reviews to a curve. Trust. When human resources, or executive management, do not trust that the review process is effectively used, then a bell-curve is employed to force the identification of excellent performance and poor performance. A bell-curve is actually nothing more than a check of data normalcy. This means that, in theory, data normally tends to migrate toward the mean, or mathematical average. This is called the Theory of Central Tendency and is a very useful descriptive statistical tool to determine if the data reflects a source of bias, or otherwise affected by a source of non-random force. When looking at an excellent example in Figure 5.4 we can see that, while the data is normally distributed it does not fit exactly to the curve, and should not be expected to do so.

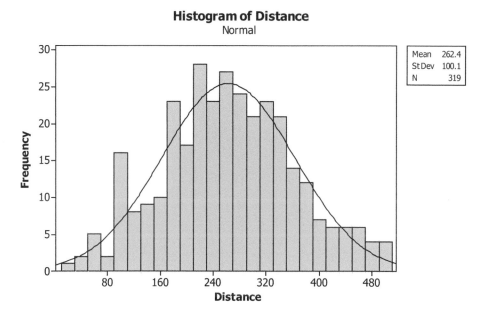

Figure 5.5   Example of a bell-curve

A normal curve is an estimation tool. When an artificial means of decision making is used such as fitting management decisions to a curve, then decisions become suspect. Managers won't effectively argue for their decisions because the decisions are not owned by them, and the problem of trust is quickly aggravated. Managers do not want to deliver a message, and they too begin to struggle with trust at an institutional level, and the problem is carried into the next goal setting period with a lack of faith in the process.

A normal curve is useful, as I said above, to find potential bias in the process. If data is skewed, or leans too heavily to the top or bottom of the curve, then human resource representatives should be asking the hard questions. Too much data at the top, and this means the managers are not wanting to deliver difficult news. This gives the leadership team somewhere to start looking for the problems, and there can be plenty of places to look. Do managers need training? Do we have a problem with trust? Should we be looking at the quality checks in the human resource processes?

PART II
Virtual Team Working

Introduction

In the past, teams have often been discussed in the light of building performance. Often authors have called this building high performance teams, and have focused the discussion on the interaction of team members with one another with the primary assumption being that team members will have ample opportunity to interact. Interaction in this context assumed collocation.

Maturity of teams working in multicultural settings has been described as a maturing process moving through the phases of forming, storming, norming, and finally performing (Shaw & Barrett-Power, 1998). Team building with the expectation that teams may move through the phases of maturity together. Shaw and Barrett-Power (1998) proposed that teams working in multicultural settings may perform well, and that performance within a team may be positively affected by the multiculturalism as they progress in maturity.

Virtuality, however, may have an effect in mitigating the ability of the team to move through the maturity stages. Elements of virtuality such as the geographic proximity to one another, use of electronically mediated communication, and a perception of homogeneity within the team are discussed in this section with the hope of providing some understanding regarding how these elements affect team trust, and therefore the formation of high performing teams. As the discussion progresses, we need to address the elements of virtuality in how teams work, and the ability of managers to recognize elements of virtuality in the ways in which their teams may be functioning.

Chapter 7, "Am I Virtual?" uses an article published in *Journal for Quality and Participation* as the basis of discussing the ways in which managers may best work with virtual teams. Managers must be able to recognize the ways in which team members interact and communicate. The office today has truly become a virtual environment extending beyond the bricks and mortar of our traditional understanding. How managers work with their teams, communicate, and motivate team members in the virtual office may be very different, requiring a focus on new skills in building trust among team members.

Team members may, as a result of the effect of the virtual office environment, react in different ways depending on the role they play in the team. Awareness on how team members in their project role react to the virtual environment may aid managers in building their team, and provide clues in how members work.

6

What Are the Elements of Virtuality?

There have been many studies and many books and articles regarding trust and virtual team performance aimed at improving virtual team formation and capability. We have often been taught how to build and improve trust in relation to team development and offshore work. Many of these publications have explored the effect of leadership, electronically mediated communications, and the need for face to face communications in order to establish and maintain trust. Until now, we have not received a lot of guidance focused on how quality systems and practices, and the roles that we play on the team affect the trust relationship.

When we choose to undertake building a virtual team, the decision to adopt a virtual setting must account for a balance between project risks, access to specialized resource needs, cost, quality, and schedule (Kahai, Sara & Kahai, 2011). These project elements are often referred to as the product value chain. As firms assess the product value chain, with current access to the technologies available today, most project resources, whether specialized skills or materials, may be accessed anywhere in the world (Kahai, Sara & Kahai, 2011). One of the necessary resources in the product value chain that may be overlooked is the need to share critical information.

Information sharing is often overlooked, not by a desire to protect information, but often by the assumption that information is readily available. Al Qahtani and Daneshgar (2010) suggest that the decision to form offshore, or virtual team, relationships must also account for the willingness and ability to share both knowledge and information.

Building strong relationships between team members requires an ability to access and share information freely, and provide the communication channels to accomplish sharing with speed and agility. This relationship and location decision, according to Al Qahtani and Daneshgar, must consider the strength of the relationships, and therefore the level of trust (2010).

As teams move to the virtual project setting, managers must make a decision as to the least risk, highest capability, and lowest cost combination of onshore and offshore, or remotely located, project resourcing plan. As Aron and Singh noted, the risk of outsourcing processes that are simple and linear handoffs between people; work that may be accomplished serially, may not provide a natural low risk option (2005). Managers need all available information regarding the best virtual team formulation in order to determine which roles may function most effectively with the lowest project risk.

Risks that are not quantified in some fashion will not be effectively factored into the virtual team formulation decision. A key risk to consider is the degree to which a manager needs to account for the potential that a virtual team setting may have a negative effect on project team member's trust for other team members. Poor trust may increase project risk; therefore it would be very useful for managers to understand if there is a significant difference in the effect of a virtual team setting on different project team members based on project team roles. This information will allow managers to identify if, when forming virtual teams, a specific role may be better positioned geographically separate from other project team members than another role, thus reducing potential project risk.

Contrasting Virtual and Face to Face Team Trust

Virtual teams often do not come about as the outcome of a planned progression of team development or project design, but rather they grow through the development and use of enabling technologies (Bergiel, Bergiel & Balsmeier, 2008). I have seen virtual teams grow by necessity into the virtual team setting as we reached out for skills across the country to folks with whom we have worked in the past.

While this may be true with traditional teams as well, the lack of personal interaction may cause virtual teams to place greater emphasis on the past experience of others when placing their trust in team members. This may even require team members to place greater trust in the judgment of their team based on the past experiences of other members of the team (Godar & Ferris, 2004). This means that one of the elements of virtuality may be the past experiences, either documented and accessed in a knowledge management database, or project closeout information that may be shared from team to team.

This database may provide the experiences of others, as well as the current team members to provide some continuity of the team experience. Members of a team may be reaching back several months, or even years to bring to the front lessons and learnings from past projects to provide context for sharing and trust on the current project.

The may be necessary due to the expectation of virtual team members' longevity. Often virtual teams come together quickly with an expectation that team membership will be short. Work will take place over a short period of time for a specific task with a very short project horizon, then on to the next engagement. Experienced team members may often be called upon to bring to the engagement a trust developed in similar settings during other team events, allowing them to ramp up with the team quickly, and come together in rapid team formation prepared to contribute.

Virtual team members (similar to that of collocated members) are likely to trust other project team members more readily when they can expect a long term relationship with future work engagements. Have you ever noticed the boldness in some folks at times when they believe they may never meet again? It seems at times that people are more candid, and ready to disagree, or perhaps defend an unpopular position, when they have no expectation that a future relationship will exist. This often means that building a strong and distant horizon in the virtual relationships may be a key element in solid virtual teams.

I have seen managers dismiss team members from the team that are working remotely swiftly at the slightest infraction of their expectations. They have never met, and therefore do not have a real face and future with which they need to feel accountable. Building a future of which the virtual team is beholden can provide needed stability for the virtual team. The expectation for future engagement can often be enhanced through contractual agreements, thus producing greater stability in levels of calculative trust (Powell, 1990).

Lacking the interaction of collocated teams, and therefore, perhaps more importantly, the interactive ability to repair trust damaged by events, temporary virtual teams may be inclined to short periods of high energy, highly interactive, high activity. This provides team members with a sense of positive reinforcement in the relationship as a means of maintaining levels of trust

(Jarvenpaa & Leidner, 1998). These self-reinforcing, high levels of activity may in part support trust in the event of low social interaction, however, Jarvenpaa and Leidner suggest, the continuous and intentional heightened project level activity enhances team confidence building between geographically separate team members. This may enhance their ability to expect continuity in project deliverables and thus their expectations for each other's abilities to be met (1998). Continuity in meeting project deliverables, as with the need to meet expectations in behavior and social norms, and process, is essential regarding virtual member allowance for risk taking and commitment to the task (Jarvenpaa & Leidner, 1998; Jarvenpaa, Shaw & Staples, 2004). This may mean that providing an environment that is rich in task descriptions and schedule milestones may be a key element in any virtual team environment.

As we build our virtual teams we must always be aware of the impression every interaction among team members may have on the relationships. In the early team building process, trust can break down rapidly as teams come together to determine the roles and pecking order amongst members. Deterioration in trust among geographically dispersed members may be aggravated when team members are located across time zones, have differences in native language, and cultural differences in team processes such as conflict resolution (Bergiel et al., 2008).

In a recent meeting designed to bring new offshore team members on board, a team with which I was working began to experience problems with the bridge line connection. Normally this may not be a problem for an experienced team, however with new members, and with each having a different native language, the garbling distortion of the phone line began to have an effect on the team. I could hear team members that were present in the room beginning to question the competence of those working remotely when they could not hear complete questions and responded inappropriately.

The remote members' ability to interpret incomplete sentences was aggravated due to the language of the conversation being their second, not their native language. The situation was aggravated by having more than two cultural groups represented on the call, and was further confounded by the faltering technology. The best thing we were able to do in the situation was to end the call with a promise to reschedule with better technology. We apologized and immediately rescheduled to prevent any further damage to the team.

Impressions, therefore, in a virtual environment, as with collocated members, must be actively managed. This means that each interaction, whether by email, video conferencing, or phone call must be consciously tailored to effect the image of a consistent, rational, group that may fulfill role requirements (Schlenker, 1975). Confidence in team members is dependent upon each interaction, and therefore each interaction should be carefully planned and executed. This may mean that communication technologies, and the effect that technology and communications may have on the ability to convey competence, is a key element in a virtual setting.

As team members we must always be aware of the effect communications may have on the perceptions of the recipient. This includes how they may perceive the unstated or underlying drivers of the communication. Dependence upon perceptions of procedural equity during times of uncertainty, according to Hakonen and Lipponen, tend to be greater in teams with greater dispersion and low co-location contact (2008). Procedural equity is the degree to which team members perceive the fair application of process when management applies decision making, and is often considered by Researches to be directly related to the development of trust and commitment. Perceptions of fairness in how group or corporate decisions are made will then have a direct effect on the degree to which team members feel an affiliation with the virtual team.

Team members must take care when relating information such as time off, vacation plans, coffee breaks, and other such subjects to team members that may be working overtime, weekends, and holidays to keep pace with the project. As noted earlier, virtual teams are often formed for short periods of high intensity work, and therefore discussions of holiday or other unplanned breaks in the work may have a strong impact on perceived equity. I saw this happen not too long ago.

The project involved several dependent projects involving teams from different areas of the company working in different geographical regions. One of the teams had fallen behind in the project schedule, and happened to be working the critical path. As a holiday weekend approached several team members participating in a bridge call began talking about their holiday plans for the long weekend. There were barbeques and family visits lined up, and a lot of jolly discussion as the week was winding down. It was unfortunate to hear the bitterness in some of the team members as they related their long weekend plans to be in the office for ten hour days all weekend.

Of course this caused some hard feelings, and several difficult management discussions regarding the need to be sensitive to the needs of team members, and their contributions. There were more than a few hard feelings following the call. Issues such as these, even when innocently raised, can create questions about the fairness of work distribution and management decisions. These simple and innocent concerns can have an impact on the level of conflict within a team, and, as Hinds and Mortensen note, such conflicts, when involving a virtual team setting, may occur more often and are often more difficult to resolve than conflicts within collocated teams (2005). Perceptions of fairness can therefore be a key contributor to trust in virtual settings.

Geography is the First Element in Virtuality

Geography, or otherwise described as geographic proximity as a variable in recent research documents, may be described as a percent of face to face opportunity in the same geographic location. Face to face opportunities are often hindered by a variety of events or circumstance such as office space or settings, corporate culture, a team's preferences for electronic communication, organizational practices, or separation by geographical locations measured in distance between physical locations (Kirkman & Mathieu, 2005).

Geography is a variable in the discussion of virtuality that shifts continuously. My team of internal consultants may, in any week, shift from collocated to virtual, to separated by miles several times a day if necessary. We meet periodically to discuss priorities, and then go our separate ways working those priorities. Geographic proximity is not always measured in relation to the degree to which mileage may affect trust and commitment.

At times the geography that separates teams is by the design of project needs. It is well understood that separation from team members, even by the preference of team members, does affect the formation of trust and the effectiveness of communication (Kirkman & Mathieu, 2005). So it stands to reason that the percent of time in which a team member may spend in close proximity with other team members does have an important effect on communications, and affects the degree to which relationships are built, and therefore how well an employee identifies themselves as a member of the team.

Since it is generally accepted that the key effect of working virtually, or in separate locations, is in the way in which separation hinders relationship building, and that poor relationships hinder the development of trust, this leads to a discussion in regard to how separation slows the development of trust. Teams need to take full advantage of opportunities to build relationships. As relationships and trust are, as we discussed earlier, about information discovery, then we will begin with the effect on our ability to help others discover information about ourselves. Self-disclosure, in the case of this discussion, may be defined as providing information in a setting in which the presenter is in control of the timing, and chooses to reveal information about personal work or life events as a means of engendering a relational transparency with a client, peer, or employee (Roberts, 2005).

As leaders we need to build in these opportunities to our work week. As I am writing today I am at the same time trying to build in an opportunity to meet informally with several peers. These peers live in a tri-state area and tend to go separate ways as soon as the end of day whistle blows. To bring the group together we are planning a one day opportunity for them to bring family and friends of their own to an offsite, informal, and fun day in the woods for picnic and play.

For me there is no better way to build in opportunities for informal talk and sharing a bit of ourselves that would not normally be seen or known to co-workers. For those that cannot meet informally away from work, and choose to use some of the earlier suggestions for off hours relationship building, it is most especially important to take care when providing disclosure as a means of creating relationships and trust. At times, the well intended transparency may contribute in creating a situation in which the person to whom the event is portrayed may misread it (Roberts, 2005). Self-disclosure can at times be misread as a form of criticism (Roberts, 2005). This may be in part due to the reality that boundaries are often very hard to find when working virtually.

Situational awareness is also a very important skill in virtual team membership. Topics of discussion and disclosure related to social events or behaviors must be carefully considered. Roberts points out that disclosure boundaries in a work setting may be different than in social settings, and the disclosure of the social self, in relation to cultural, sexual orientation, ethnic identity, and so on, may be considered by some team members to be inappropriate (2005).

We each present to others a personality, or face, that we wish them to know (Altman & Haythorn, 1965). The face that we present, or the self we wish the audience to know, may necessarily change based on the setting and likely the timing of a disclosure event as well (Altman & Haythorn, 1965). The person or the self we present in the work place may change with situational dependency, shifting our self-identity based on the setting and receiver of the intended or unintended disclosure (Altman & Haythorn, 1965; Roberts, 2005). Disclosure of one's personal feelings or other information which may carry great personal meaning, in the work place, requires a constant weighing of appropriateness, and what may be useful to disclose in a specific setting or situation.

The process requires a clear understanding of the social boundaries present in the situation which may be hard to discern in a geographically dispersed work team. Through this process of boundary identification, both internally and externally, the process of self-disclosure requires the determination of an acceptable social level of disclosure and extended vulnerability acceptable to extend the trust relationship (Watkins-Allen, Armstrong, Riemenschneider & Reid, 2006). Social acceptability becomes complicated in the absence of the social clues normally present when collocated.

Although separated by miles, or merely by corporate practices, teams are now often connected 24/7 by text, telephone, wiki, email, blog, social networking sites, and virtual game worlds connecting workers in ways never before considered. When supervisors enact procedures that allow for greater interaction between separated team members, and provide newer employees opportunity to interact with those having more experience, team performance is often enhanced as trust strengthens.

Almost 40 years ago Cooper lamented the lack of computer knowledge in software development managers (1978). Today our world is one that, much like the challenges technology created in 1978, managers must learn to exploit and master the generational differences as companies scramble to learn to connect the incumbent employee with the social practices of the young and coming new worker. This may be particularly important in developing trust in newer employees.

The new employee begins their work relationships often feeling isolated and at risk, and with the rate of change in the corporate environment today, employees will often feel a heightened level of risk. With change of any kind comes stress, and when adding in the geographical challenges in isolation of the modern workforce and the gaps in information this may create, stress may invoke the tendency to project unreasonable meaning into the gaps.

Workers in isolation, or feeling isolated due to separation from team members, may fill information gaps with thoughts of possible future organizational changes, or potential future decisions of organizational function or structure, and other future work events, postulated from present conditions in an attempt to create personal context where formal context may not exist (Fiske, 1993; Schwarz & Watson, 2005).

Gibson and Hodgetts remind us that well formed and planned disclosure of self in the work place may reduce the perception that managers are set apart from the work force, increase communication and sharing of scarce information and open up new avenues of feedback (1985). Management relationships, either formal mentor–mentee relationships, or simply paired workforce training, require open communications, active listening, and a solid level of self-disclosure fostering high degrees of trust to support the necessary risk taking for employee growth (Hale, 2000; Liu & Batt, 2010). Within cross-cultural, or simply geographically dispersed relationships, in order to bolster what tend to be weak relational links, the participants should assess their ability to communicate, and the style that is often used in communicating information.

There is a relationship, Wilkens and London noted, between reduced team member conflict, and the degree to which team members feel safe within the team setting (2006). Those teams that perceive higher levels of self-disclosure also perceived lower levels of conflict and vulnerability (Wilkens & London, 2006). This, according to the authors, was reflected in higher levels of reported trust.

Communication and Virtuality

Communications is defined as the transmission or reception of information or knowledge, either face to face, or through a mediating technology, and is often measured as a percent of time that team members communicate one to one. Communication is often considered to be an indicator of team virtuality which may be estimated as a percent of a team member's communication with other project team members that is accomplished using electronic media in a typical week. With strong and effective communications considered to be an essential element for every team and the formation of trust, then the level of communication accomplished using electronic means may be considered to be a key variable in virtuality.

Closed loop communications are often considered by many experts to be a measure of communication effectiveness, and also to contribute to greater levels of performance in the areas of creative problem solving and team process changes. It is commonly accepted that teams in a virtual setting are beset by communications challenges, many of which may hamper the development of a consistent and open environment, and the expression of self-disclosed events that are necessary for the development of trust. Receiving feedback in the conversation, or reciprocation in self-disclosure events, as well as the regular feedback necessary for effective leadership, is commonly recognized to be significantly correlated with increases in team performance.

Another attribute commonly accepted as a measure of effectiveness in communication is the degree to which non-verbal signals are received and processed. How information is conveyed, though, depends greatly on the degree to which the sender depends on non-verbal signals, and the circumstance in which these signals are used. We'll talk a little more about non-verbal signals as this discussion progresses.

I'll never forget the day I was called into the CTO's office to explain why I felt empowered to make demands on his time. With great wonder and trepidation I was summoned to his office to explain myself, and so I immediately set off to find out what I had done. When I arrived, he brought an email to the front of his computer desktop, and while pointing angrily at the screen demanded to know why I was making so many demands.

Shaking my head with wonder I again asked what he meant. He pointed to his screen again telling me to focus and explain my email as he jabbed an angry finger at the underlined sections of my email. Finally getting it, I asked why he felt these were demands. They're underlined he exclaimed. Underline means that you demand.

I explained to him, that underline meant that I realized he would be using a handheld device to read his email while traveling, and that I wanted to be sure he knew which part of the email was important. I would never demand anything from a chief level leader, and of course if I did I would not do it by email. Just as a quick note, we all have to be very careful with electronic communications. The CTO's level of trust had dropped a few notches that day as he sent me away with an assignment to learn email etiquette.

As we all learn to communicate effectively in support of trust, and therefore grow a good healthy relationship with team members, a sense of connectedness may be created; sometimes described as a perception of family (Halgin, 2009). Family is described by Halgin as an enduring relationship and connectedness that carries with it the expectation that the relationship will last into the future, with future connections effecting trust in coming engagements (2009). The perception of connectedness, or that point at which an employee identifies their self as having membership in the organization, relates to the earlier point of self-categorization. As a team member identifies themselves as a member of a team, this increases the sense and perception of belonging and enhances their trust in the team lowering their sense of vulnerability and enabling self-disclosure opportunities and willingness (Elving, 2005; Haslam, Postmes & Ellemers, 2003).

Polzer et al. suggest that the process in which one's connectedness develops, or as Polzer, et al. describe the condition of connectedness as group reputational validation, may take time, involving give and take within the group, or reciprocation regarding one's positioning among group members (2002). The process of building trust through communication requires feedback, and as Polzer et al. (2002) note, feedback may not always need to be positive to reinforce the process of disclosure and reciprocation.

A simple example of needed negative feedback may be my very inept ability to match ties, shirts, and suits, and the reliance I have on open and honest feedback to avoid embarrassment. I count on others to tell me if my clothes match. If I somehow find myself in public, and on display with my poor sense of style, and then realize people had an opportunity to provide feedback and chose not to share this bit of information, my ability to trust in them may be diminished.

In building effective, high performing teams, feedback as well as self-disclosure is necessary; both positive and negative. The process of self-disclosure and feedback is often more complex in a virtual team setting. We can, however, reduce the effect of distance on communications through the use of electronics and leadership motivational techniques.

Moore (2007) found that leadership behavior does have a direct correlative relationship with virtual team member motivation. How leaders conduct standard behaviors such as conflict resolution, coaching, goal setting, and specific attention provision, as well as the ability to provide interesting work, does have an effect on virtual team motivation and performance. Moore discovered that as much as 42 percent of the variation in employee motivation may be accounted for in leadership behaviors (2007).

Trust is often hampered when effective communication practices are not used. Issues such as interpreting, long periods of silence, and an uneven distribution of information due to a lack of response can challenge the trust relationship. Where trust exists so may acceptance by one's peers, and openness of self-expression, supporting the opportunity for vulnerability and increased self-disclosure for relationship building. Maintaining a level of perceived positive reciprocation of communication and self-disclosure is necessary to maintain the level of closeness obtained in the relationship.

Communication is an essential element to effective team function. As a manager, however, we need to be aware that communication should not be assumed to be effective within virtual teams since the common cues normally assumed as a social function for understanding are often not available to a virtual organization. Thus the management of communication mediation must be a high priority in the establishment of teams and team formation. While, as previously stated, electronically mediated communications are an accepted means of communication, they are sometimes considered to be depersonalizing. This is because some believe that reliance upon electronic communications can lead to a culture that is comfortable breaching social norms and boundaries, and weakening the normally present social controls enforced in face to face communication (Postmes, Spears & Lea, 2002).

Otherwise said, people may be perceived as being too bold, or as rude, when using electronic communications. The use of electronically mediated communication may also be perceived as increasing the separation between groups in the work place. Participants in electronic communications may tend to carefully evaluate the non-verbal signals in this type of communication in an attempt to glean the full meaning, thus hampering a potential trust building opportunity. If we accept that building a trust relationship may be cumulative in nature, then the absence of social norms in mediated communications may create some challenges in increased barriers boundaries between work teams.

Elving notes that people have a tendency toward social categorization (2005), The process of self-categorization, or the development of one's identity as a member of a group, occurs as an employee begins to categorize themselves as belonging to a work team, or as a member of the corporation in which they work (Elving, 2005). Communication is believed to play a strong part in the formulation of the self-categorization process in which a team relationship and identity is established. The extent to which an employee perceives themselves to be a part of an organization, group, or team, may be strongly influenced by the level of emotional attachment, which in turn may be influenced by the communication strategy.

Commitment to the work group, Postmes et al. states, may be an outcome, or even synonymous with, the emotional attachment, or the degree to which the employee maintains self-identification with the group (2001). Some research has shown that an employee's commitment to the work group is strongly related to the common issues of absenteeism and attrition in the work place. Therefore, the commitment, or level of attachment to the other team members, has a high potential to affect the attainment of team goals.

Socialization through effective communication is necessary for team members to build a team identity and form that important bond of concern among members to strengthen trust (Bhattacharya, Devinney & Pillutla, 1998). It is well established, and commonly accepted that, team members that work quietly in isolation from the team, working in a self-directed manner as virtual team members may do, will often experience lower levels of trust and team commitment. This may be manifested in behaviors that are contrary to the goals established by the team. At times, isolated teams have been observed to display higher levels of behavior contrary to the best interest of the team, and seek out greater isolation from other team members (Godar & Ferris, 2004).

I had an employee long ago when I led a team of technical writers that had a tendency to work quietly, and prefer isolation. This fellow would even go so far as to find a place under the stairs or in storage in an attempt to isolate himself from contact with other team members so that he would be able to work alone. While this may seem odd, it is not necessarily. Writers, like developers, require long periods of uninterrupted thought, and when working in loud or crowded environments, the interruptions that come with contact with others can create a large amount of restarting and reorienting to the work product.

He would find ingenious ways to hide and work. This was misinterpreted by some as a signal that he did not want to be a part of the team, which tended to cause anxiety among the team, and moved some to parse his email and phone messages with great detail and consternation, driving him further from them. This caused him to begin to have trouble getting work through his support teams such as word processing and the review and approval process. When this happened he began to move into greater isolation and eventually began working outside of the approved processes to get things done, creating problems for the entire team that included questions from the union representatives, and a great deal of extra work for the management team to settle questions with our union.

It took a lot of effort to bring this person back to the fold. One of the processes that we used was setting a pattern to communications that included meeting face to face on a regular basis. This can be done in person, or with the aid of electronic video opportunities, but however it is accomplished, it goes a long way in heading off team member isolation; even when the isolation may be a preferred condition. Face to face, however, does not automatically imply co-location. The use of electronically mediated communication such as video conferencing, computer based video calls, or other forms of visual contact such as smart phones, work nicely as well.

Culture and Virtuality

The success of self-disclosure events in the virtual work setting is often dependent upon the contractual agreement between organizations, and cultural differences between participants (Horenstein & Downey, 2003; Panteli & Duncan, 2004). The rate at which the self-disclosure events are offered are often different between cultural groups, and are often based on cultural issues such as the level of collectivism, and yet it seems to be common across most cultures that the rate of reciprocation does affect a positive pressure on openness. This is possibly due to the need for reciprocation in self-disclosure to precipitate the extending of trust. Culture, in a cross-organizational environment such as outsourcing, may play a strong role in developing communications and effective virtual teams.

Cozby noted that the level of self-disclosure received is correlated with the level of disclosure provided (1973). Maintaining a positive level of reciprocation in communication and during self-disclosure opportunities is necessary to maintain the level of closeness obtained in a work relationship. When we work with cross-cultural teams, we should expect that trust can be hampered due to communication issues such as interpreting a lack of response, low levels of reciprocation, long periods of silence, and uneven distribution of information (Cozby, 1973; Powell, Piccoli & Ives, 2004). To better manage these issues we as managers should seek to staff teams that will be working with other cultures with those team members with more experience working internationally.

Team members that have the experience of working with other cultures are generally more willing to openly discuss and accept topics related to those of other cultures, and therefore often more capable of self-disclosure in a multicultural environment, thus more capable of engendering trust and commitment. To assist teams in being open to talking about subjects outside of the immediate work topics, managers need to ensure that team members understand and are comfortable with issues such as non-disclosure agreements and the opportunities that these agreements provide in freeing up discussions.

Disclosure of one's self in the virtual work setting is often dependent upon the contractual agreement between organizations, as well as cultural differences between participants. We know that rates of self-disclosure are often different between cultures, as well as between males and females within cultures, and are often based on cultural issues such as the level of collectivism. Some researchers have discovered, however, that openness can be improved when the discussions are shifted to the use of electronically mediated communication.

7

Am I Virtual?[1]

Virtuality is not measured in distance, through electronic communications, or as a measure of cultural homogeneity. It is a critique on how work gets done. While each of these factors impact the way we work, virtuality is in reality affected by the factors as they affect trust. Managers often experience co-workers sitting on the other side of a cubicle wall working silently, and separate, and yet together on a project deliverable. Silence can invade the work place as a new generation of workers hunkers down with their nose to a virtual grind stone.

Virtual teams are often described in literature, and studied in recent research, as offshore teams brought together by managers for a short period of rapid activity designed to focus on a well defined goal. The Fairfax County Virginia Economic Development Authority quotes a recent report projecting that almost 25 percent of the US workforce will be considered contingent employees by the end of 2012 (April 2012). In this report, a contingent employee is described as temporary help, part time help, contractors, and virtual employees (Fairfax, April 2012).

Teams of employees are created of folks that likely have never worked together, have never met, and increasingly may have been hired virtually, meaning they have never met their employer face to face, but were hired through the internet (Fairfax, April 2012). The US Department of Labor Statistics has named the trend toward virtual work as virtual immigration noting an increase of over 180 percent over the ten year period between 1998 and 2008 (United States, February 2010). These team members are dependent upon the capability provided in electronic communications and virtual private networks (VPN) to bridge gaps in both geographic separation of team members, and to leap time zones. Truly virtual team members with the technical savvy and

---

global experience hired to join teams as virtual members are often relied upon to have a greater capacity to rely on trust developed through institutional engagements.

In our case, a virtual team is any team that relies upon electronic communications and technical media as the primary means of communications and data access, and does not regularly meet face to face. There are very good reasons why a definition of a virtual team or group must be expanded. Team members may work in the same building for years, and never discover each other, or for that matter even to attempt the discovery, and yet will have enjoyed working with many folks across the globe that they may call friends and co-workers.

A few years back there was a popular and simple parlor game people enjoyed playing that is called 'Six Degrees From Kevin Bacon.' In this game players were challenged to take any Hollywood movie production, and through the names of the actors and other movies in which the actors worked, build a logical connection to a movie in which Kevin Bacon had appeared. The game was a popular way of bringing realization to the public that the world in which we live and work truly depended upon the small world concept.

The idea of the small world came to the world consciousness in the late 1960s which described the concept of two random people that perhaps would meet in a diner far from their home, and somehow discover they have a mutual friend or acquaintance. I experienced this only a few weeks back when my family and I visited the Kemah Boardwalk in Houston, Texas. Kemah is a historic little town more than 1,500 miles from home sitting on Galveston Bay.

As my wife and children and I were leaving the boardwalk after a beautiful evening on the bay, and casually strolling to our rented minivan we continued to cross paths with another group of weary travelers. Upon the third time we crossed this group I decided to strike up a conversation. I said hello as we met again working our way to the parking lot, and offered up a small conversation. As we talked I discovered that he lived in the same town from which my family and I had moved five years back. We had shopped in the same stores, and knew the same store keeper.

I had this same experience a few times as a child in Virginia when I met a boy on my baseball team with whom I shared a great aunt from Illinois, and another time as I played in our yard and spotted a boy with whom I shared a classroom in a California grade school now walking down the sidewalk in front of my Virginia home. In a small world these experiences seem to appear in our lives periodically,

I had a similar experience as a young adult on vacation in Florida. As I walked across the main square in Disney Land in Orlando, Florida I heard my name yelled from across the square. A friend from my high school back in Joliet, Illinois that I had not seen in more than 15 years was frantically waving to get my attention. We had a wonderful time sharing memories, and then went our own ways once again. I have not seen him since that day in Florida, but I assume some day we may meet again in a faraway restaurant perhaps.

There are no longer 6 degrees of separation from Kevin Bacon as social media continues to have a dramatic impact on our small world. In 2008 it was reported that 64 percent of teens engage in some form of electronic mediated communication (Lenhart, 2009). Often organizations as well as individuals will use social media to instantly express a political view to influence the masses before investigation of events by the established reporting of the events can take place.

Flash mobs and flash Christmas caroling at the shopping mall, government collapse in the middle east, and fan morale collapse as NFL players tweet their dissatisfaction with the Chicago Bears quarter back as he walks off the field during the 2011 NFC Championship game, are all derivatives of the new social media power. Corporations use social media as a means for communicating executive sentiments. Text messages are often the favored off hours communication tools for many socially savvy team members to express snippets of concern regarding after hours work.

Relationships and expectations are rapidly evolving in an always-on-world (Kim, 2000). Each of us has probably experienced the site of an employee texting another across the room, or from cube to cube, rather than walking over for a brief moment of face to face communication. Each of us has further experienced texting a child in another room of the house as the best means to gain their attention. The world is no longer small, but is now very small and very flat, as everyone is now accessible and total strangers are now closely trusted with the most delicate feelings (Gilbert & Karahalios, 2009).

Work place performance is affected by access to social media, both in positive as well as negative ways. I recently met with co-workers from another work group. Folks that I had never met prior to this meeting, and may not meet again face to face for some time were in attendance, and yet were able to ask questions regarding my past experiences that would not have been apparent to them had social media not presented the opportunity. A brief search on the web provided them with my past work experiences and projects, and seeded the conversation with only a little effort in advance of the meeting.

On one project I worked a few years back, team members were caused to suddenly depart the work place as a result of too much reliance upon electronically mediated communications in a manner that became spitefully dependent. A refusal to discuss any project subject by any means other than computer aided chats started to create long project delays. This happened when one team's access to the project online chat failed due to a glitch in the system, while the other team continued to send silent and unanswered queries of "anyone out there?" The silence on the other end was interpreted as an absence, rather than a failed communication while project deliverables began to slip.

One question that is often raised is how communications through modern electronic media may affect employee performance and project outcomes when conditions or practices preclude the common face to face communications. As electronically mediated communications invade the work place, and more companies make the move to virtual project teams, the question of motivation and work performance becomes more complicated. Performance and motivation are closely linked through literature and practice, and one cannot be effectively discussed or managed in isolation. This chapter explores the question of motivation and employee performance in relation to the use of electronic communications media and virtual work teams.

Performance, according to Whetten and Cameron, is a quotient of ability and motivation, and ability the quotient of aptitude, training, and resources, while motivation is the quotient of desire and commitment (Whetten & Cameron, 1995). These seven variables come together to make the backbone of every project team with the capability of impacting a team's success or failure. In relation to our question regarding the effect of modern electronic communications on project team performance, managers may seek to address the motivational factors in Whetten and Cameron's performance equation.

Maslow provides motivational theory in a graphical representation of unsatisfied needs in a hierarchical pyramid based on the human desire to satisfy those needs of the most basic level prior to graduation up the hierarchy

(Maslow, 1948). Physiological and safety needs, according to Maslow, must be satisfied before the employee is driven toward satisfaction of needs at the social and self-esteem levels (Maslow, 1948). At the pinnacle of Maslow's hierarchy of motivational needs is the desire for self-actualization as the employee seeks to reach their full potential driven by intrinsic needs for satisfaction, as the employee's needs in health, safety, bonding, and respect are met.

Herzberg compliments Maslow with the theory of hygiene factors in motivational theory seeking to explain the de-motivation, or stepping down Maslow's hierarchy, as lower level needs are raised regarding self-esteem, and social needs. These social needs may be met by the team as team members identify themselves as members, allowing them to find safety in team membership. Communication practices and strategies allowing for effective trust development, therefore, may be able to assist in effectively allowing for greater motivational practices within the team environment.

Maslow and Herzberg form a basis for further understanding of employee motivational research in the twenty-first century as managers seek to understand motivational theory in relation to project teams. Blaskova explains that motivation is formed intrinsically through a workers internal belief system (2009). The worker may extend trust through a predetermined response based on expectations, known as *institutional trust*, which is dependent upon the degree to which the worker perceives the management system to be consistent and equitable.

Workers are often willing to adjust their performance based on the presence of motivational factors in work place (Blaskova, 2009). Maintaining a motivational work place, however, is becoming more difficult as teams become increasingly dependent on electronic communications and are dispersed throughout the world raising concerns at the lower Social and Safety Needs of Maslow's pyramid. Powell expressed this concern declaring that as teams move to a virtual team model, whether by corporate design or as an expression of social and generational difference, socialization of the team may be reduced, hindering the establishment of support and the membership employees once gained through work place relationships (2000).

Powell, Piccoli and Ives questions how, as team members increasingly never meet in the break room, often don't verbally communicate or commune in the hall way, many times have never even met, can the social bonds in Maslow's hierarchy effectively form (2004)? Weems-Landingham, writing in 2004, distinguishes the natural difference between virtual teams and traditional teams as not simply the physical separation, but the additional challenge of the psychological dispersion. The psychology of the geographic separation can, at times, be an unkown–unkown in the project risk assessment that many modern managers are not equipped to handle.

Yet is this merely a generational question for today's management team? Writing in 1978 Cooper lamented the lack of development knowledge in the managers assigned to develop software, and yet today it is hard to find a young person without some level of electronics expertise. Although separated by geographic miles, or merely by social practices, teams are now connected 24/7 by text, telephone, wiki, email, blog, social networking sites, and virtual game worlds connecting workers in ways never before possible. This world is one that, much like the struggles of management in 1978, managers must learn to exploit and master the generational differences as companies scramble to learn to connect the incumbent employee with the social practices of the young and coming new worker. As Kaplan and Haenlein note, the practice of social media is now a revolutionary trend, pulling the world together in ways baby boomers could never imagine, changing the psychology of team socialization (2010).

Moore addresses the issue of the psychology of team socialization indicating the need for leadership motivational practices, such as goal setting, personnel development, and activity coordination, are absolutely essential in strengthening virtual and geographically dispersed teams (2007). Having to work through technologies such as web messaging, text, email, and other forms of communication interventions, Moore suggests that team leaders be required to have the skill sets necessary to communicate in a twenty-first century environment to build the relationships between team members, create the psychologically safe work environment, and provide encouragement and opportunity for professional growth (2007).

Team socialization is a required motivational tool that must be provided to address the needs that Maslow described for social level actualization before a team can attain greater satisfaction and motivation. Managers of effective virtual teams, Weems-Landingham says, rely heavily on understanding the feelings, work problems, issues, and other situational conditions, as well as individual motivations and team member interdependencies to demonstrate situational awareness and empathy for team members (2004). Additionally, the ability to listen attentively and actively in a non-evaluative way is heightened due to the absence of face to face communications.

Understanding feelings, listening, solving problems, and situational awareness require the ability to discover relevant information. This is the key message in this book, and managers must be capable of understanding the information and asking the right questions. Therefore the manager of virtual work teams must enlist every available means in the new world of electronic communications to listen and gather facts in order to fully illustrate an understanding of the factual conditions, as well as employ the traditional methods of listening, empathy, teambuilding, and participation to bridge the generational or geographic gap (Weems-Landingham, 2004). Text messaging, instant messaging (IM), email, and other electronic media are effective and essential communications devices and strategies for effective work teams and should not be a disruptive or limiting agent in communication and team socialization. The effective manager of high performing teams must find creative ways to fill the needs of the employee in safety and socialization if the team is to be a truly effective virtual workforce.

A New Office Communication Environment

Communication is the first factor we normally think of as affecting virtuality, and for good reason. Communication is essential to business, community, and a sense of well being. Communication is an essential element in diplomacy (Jönsson & Hall, 2002). Since Merriam-Webster defines diplomacy as having skill in, and without arousing hostility in, the handling of one's affairs, we might then conclude that communication is the essential element necessary in handling our affairs in community with one another (2010).

As I walk around the office environment it is easy to see many different forms of communication. Inevitably we find folks standing around an office desk chatting about last night's sports scores. However, when we take a closer look, it is highly likely at least one of the chatters has a smart phone in hand texting a virtual participant, and checking a web site as a way of verifying the score of the game to support their argument.

Their discussion is likely to shift to work after a few moments enjoying and rehashing embarrassing moments for their favorite athlete. The other participant in the discussion will grab their smart phone to check on a scheduled meeting time, while the other will quickly IM a colleague to check if they are online, while quickly popping up their favorite social networking site. On the site they may be checking on the progress of a club, or perhaps a graduate school project. Seventy percent of adults are now using social network sites to stay in touch (Lenhart, 2009).

Electronically mediated communications is often the chosen way in which team members communicate. In a recent study on the practices of project team members, questions regarding the use of electronic mediated communications were presented to the study participants to identify the percent of communication with other project team members accomplished using electronic media in a typical week (Wise, 2012). Electronic communication is reflected as the prominent means of communication by respondents with 69.89 percent of respondents reporting 50 percent or more of communications by electronically mediated means. See Table 7.1.

Table 7.1   Respondents reported time in electronically mediated communication

Amount of Electronic Communication				
	Count	**Percent**	**CumCnt**	**CumPct**
0% to 24%	67	11.39	67	11.39
25% to 49%	110	18.71	177	30.1
50% to 74%	214	36.39	391	66.5
75% to 100%	197	33.5	588	100

Note. The column titled CumCnt expresses the cumulative count using the data in the column titled Count. The column titled CumPct expresses the cumulative percent using the data in the column titled Percent.

As more employees embrace electronic communications we have to acknowledge the need to accommodate those needs in our quality and communications programs. As generation X takes the helm in many corporations, and Millennials move into the work force in greater numbers, the use of electronic communications is increasing. As the New Silent Generation comes of age the use of online communication, text messaging, IM, and social networking may be the predominant communication device.

The bottom line, if there are more than one of you in the office, or working on a project, and you have some reason to communicate with someone else, to a great extent, your company is at least in some fashion virtual.

Geographic Proximity May Have an Affect

Geographic proximity raises the question of whether team members, or employees with shared goals, are collocated. This is really a question regarding access to one another in a face to face format. While it is clear from the previous discussion regarding communication styles and habits that co-location is not a resolution to the problems raised by virtuality, it does have an impact on how people perceive their relationship with co-workers.

Geographic proximity is merely a fancy way of defining the location of employees, co-workers, and team members in relation to each other. Collocated can mean working in the same state, town, or building, and is measured as a percent of work time in close proximity with team members with whom the respondent to the question is actively working. If employees can come together physically whenever the need arises, and communications are therefore face to face when desired, and at a cadence that supports information discovery and self-disclosure, the team members may be considered to be collocated.

Millward, Haslam, and Postmes noted in 2007 that hot-desking, or hotel seating as some companies have named the practice, does not necessarily affect negatively an employee's perceived membership with company. It can, however, have an effect on how they perceive their membership within the work team. According to Millward et al., those employees that are not assigned a seating location within the work group tended to pull away physically from the team, and were more likely to work with other team members using virtual means (2007). The real measure of co-location to answer is in regard to the amount of time team members are able to work or meet face to face when working together on a project or shared goal.

When we look at the way teams are formed and work today, it is quite different than when I first arrived in the workforce. A recent study shows that 44 percent of a given sample of workers spend more than 50 percent of their time working in the same location as their project team members. Forty four percent of their time on a project collocated may sound good, but the absence can take a toll on productivity due to the way it affects perceptions of team membership.

On the flip side of that study is the reality that more than 56.29 percent of respondents spend less than half of their project time in face to face, or in collocated presence. See Figure 7.2. The middle two brackets representing

25 percent to 74 percent of project time spent face to face represents 48.81 percent, or very nearly half of the respondents.

Table 7.2   Respondents reported time working in the same physical location

Time in Same Physical Location				
	Count	Percent	CumCnt	CumPct
0% to 24%	198	33.67	198	33.67
25% to 49%	133	22.62	331	56.29
50% to 74%	154	26.19	485	82.48
75% to 100%	103	17.52	588	100

Note. The column titled CumCnt expresses the cumulative count using the data in the column titled Count. The column titled CumPct expresses the cumulative percent using the data in the column titled Percent.

As we discussed earlier, working collocated does not automatically translate to a non-virtual environment, however it does help. As the time spent collocated with other team members decreases, the time spent in electronically mediated communications increases. Naturally, those that spend more time collocated will spend less time communicating by electronically mediated means, therefore potentially increasing the likelihood of developing stronger bonds and work relationships. See Table 7.3.

Table 7.3   Respondent time spent in electronic mediated communication is inversely affected by time spent co-located

Electronic Mediated Communication	0–24% Time Co-Located	25–49% Time Co-Located	50–74% Time Co-Located	75–100% Time Co-Located
0% to 24%	17.65	15.69	18.81	50.47
25% to 49%	11.76	19.61	29.7	11.21
50% to 74%	23.53	35.29	32.67	19.63
75% to 100%	47.06	29.41	18.81	18.69

Rapid Trust Development May Be Needed

Swift trust, commonly described as an assumed trusted relationship, is often present in project participants with experience working in virtual teams. Having a strong background in working with virtual project teams is often considered by some to be essential in development of swift trust, however the development of swift trust may be enhanced by contractual agreements that support the mutual sharing of information. These agreements are common, and are designed to allow team members to share information and speak without fear of disclosing protected information, however they can also provide some comfort to teams that a longer term relationship is possible and can therefore produce greater expectation of future work together creating an increased level of calculative trust.

So what does that all mean to me? If projects need to be built on virtual teams, the greater the experience level of the team members in working with virtual team members, the more likely the team will be able to ramp up quickly in productivity. Companies that specialize in offshore work arrangements, and team members with a lot of experience working remotely are often able to make the assumption of a trusting relationship, and begin the engagement ready for the hard work more quickly.

The Look and Feel of the Office is Different

Way back in the 1990s the culture of the office environment began to change. It shifted from being a place that everyone looked and talked the same, to a place where the middle class white male no longer dominated. In a recent labor force report by the United States Department of Labor, experts project that over the next decade the US labor force will continue to grow more ethnically and racially diverse (United States Department of Labor, 2012). Over the next decade, 2010 through 2020, the number of white non-Hispanic workers is expected to drop approximately one and half percentage points as a percent of the total work force (United States Department of Labor, June 2011). At the same time Asian and Hispanic workers are projected to increase approximately 30 percent, continuing to change the office demographics, while 42 percent of the management and professional workers in the US in 2010 were foreign born workers (United States Department of Labor, February 2012; United States Department of Labor, June 2011).

This shift from homogeneity began to happen, not just in the American office, but expanded and accelerated over the world as businesses began to embrace the virtual world of the internet, and we all became more aware of each other's contribution. As we came to know and accept one another, we became more acceptant of our differences and learned to capitalize on our differences as well as similarities.

Culture, or homogeneity, is measured in the degree to which we find our team members and co-workers to be similar to ourselves. It is interesting to find a recent study that 49.83 percent of respondents perceive their project team as mostly homogenous. Only 15.14 percent of respondents reported a greater than 75 percent or more homogeneity, and a corresponding 15.65 percent of respondents report a 24 percent or less homogeneity in the project team (Wise, 2012).

A majority, or 69.21 percent, of respondents reported homogeneity, or likeness within the team, as between 25 percent to 74 percent (Wise, 2012). See Table 7.4. So, it may be fair to note that, although the world of business is very mobile, and in most cases internationally connected, we still perceive the similarities among us quite clearly. Additionally, this study does show that most team members and co-workers work in the grey zone of virtuality where the need to ensure greater information discovery may play an even stronger role.

Table 7.4   **Respondents reported perceived homogeneity of the project team**

Perceived Homogeneity of Project Team				
	Count	Percent	CumCnt	CumPct
0% to 24%	92	15.65	92	15.65
25% to 49%	203	34.52	295	50.17
50% to 74%	204	34.69	499	84.86
75% to 100%	89	15.14	588	100

Note. The column titled CumCnt expresses the cumulative count using the data in the column titled Count. The column titled CumPct expresses the cumulative percent using the data in the column titled Percent.

Culture plays a very small role regarding an effect on our ability to perceive trustworthiness in our relationships with team members (Wise, 2012). Personality based trust does not appear to be affected by the degree to which team members and co-workers perceive the team to be homogenous, however the effect is significant when we talk about institutional and cognitive based trust. As homogeneity is perceived as less within the team, the level of cognitive based trust and institutional based trust of other team members that play different roles may begin to drop (Wise, 2012).

Additionally, project teams may have many cultures, many roles, and many personalities represented with team member contributions varying in complexity and size, as well as priority, and with these variables potentially may arise conflict and questions.

Conflict and Uncertainty

Teams distributed geographically may tend to experience conflict differently, and conflict may be engendered differently than experienced by collocated teams (Hinds & Bailey, 2003). Distance and the use of mediated communications may, according to Hinds and Bailey, give rise to conflict simply due to these two conditions (2003). This may be due, according to Armstrong and Cole, to the length of time it takes to identify and address the conflict when working separate, yet together, on a project (2002). Uncertainty may create not only fear but ambiguity, the seriousness as well as the number of questions regarding priority and methods creating a heightened risk of delay.

Managers seeking to ensure an efficient and effective project team need to be able to manage the questions and fears that arise, and prevent the unnecessary interruptions to project delivery that may come with them. Establishing an open and trust supporting relationship, and the virtual work environment that creates trust, is critical in mitigating project risks.

Many elements of project work may mediate the rise and management of conflict in geographically separate work teams. Purdy, Nye, and Balakrishnan noted that the ability to work collaboratively may be reduced when working geographically separate, and therefore it is less likely that the team may find a collaborative resolution to conflicts (2002). Task interdependency is a primary example of conflict moderators. This means that as tasks dependencies increase, the likelihood of conflict may increase as well. Other moderators may include the number of people on the project and the cultural make up of the team.

Teams must readily learn from their experience when working in a virtual environment, and be able to collectively adapt their use of technology and practices from these experiences to aid the team in resolving work practices to the benefit of the team. Learning is often an exercise in trust and an improvement of personal and work relationships developed between team members. This requires, as suggested earlier, a good amount of relationship building, which in turn requires a good amount of self-disclosure.

Sternberg and Grigorenko (1993) noted that task interdependency may be related to conflict in virtual team environments. Members need to have clearly defined roles and responsibilities that hold autonomy from other members (Sternberg & Grigorenko, 1993). Reducing task overlap may serve to reduce opportunity for conflict (Sternberg & Grigorenko, 1993). As team members gain experience and problem solving skills rise, and as trust may therefore begin to increase, commitment to the team may also increase improving the ability of the team to reach their project goals (Park, Henkin & Egley, 2005).

The conflict resolution and avoidance needs may be a bit more complicated in the use of newer, less process oriented project methods such as in the use of agile development. Member roles, when working with agile teams, may intentionally remain unclarified. Multitalented individuals may be able to perform multiple roles, and assignments may be picked up by team members focused on the goal rather than the role (Larman & Vodde, 2008). This may, however, be a matter of awareness rather than an active management task.

In many cases it is not unusual for teams to have members working in several roles. Table 7.5 reflects data collected in a recent study that shows 75 percent of the respondents reported participating in projects in the role of tester, with 13 percent of those respondents reporting a tester role also reporting 50 percent or greater time spent on projects in the tester role. As this table reflects, most respondents report working in more than one role on their most recent project. With this aspect of project work so prevalent, the issue of separation of tasks may increase the possibility of conflict.

Table 7.5   Frequency of employee roles reported by respondents

Frequency of Project Roles						
	Tester	Developer	Analyst	Engineer	Project Manager	Other
0% to 24%	337	307	226	245	70	206
25% to 49%	47	48	143	95	107	59
50% to 74%	27	13	54	29	159	30
75% to 100%	32	13	14	11	182	34

Project manager is the most often reported role at 88 percent of respondents, which in turn reflects a high profile role that may potentially lead to conflict due to multiple role responsibilities (see Table 7.6). Only 74 respondents, or 13 percent of respondents, report serving exclusively in one role.

Table 7.6   Percent of employee roles reported by respondents

Frequency of Project Roles						
	Tester	Developer	Analyst	Engineer	Project Manager	Other
Percent respondents reporting each role	75%	65%	74%	65%	88%	56%

Sixty two percent of respondents reporting an exclusive role in one area reported that role as project manager, and 24 percent of those reporting an exclusive role report that role as tester (see Table 7.7).

Table 7.7   Respondents reporting exclusive roles

Respondents in Exclusive Roles						
	Tester	Developer	Analyst	Engineer	Project Manager	Other
Number Respondents	18	2	0	0	46	8
Percent Exclusive Respondents	24.3%	2.7%	0.0%	0.0%	62.2%	10.8%
Percent Overall Respondents	3.1%	0.3%	0.0%	0.0%	7.8%	1.4%

So what do all of these stats really mean? We discussed earlier the prevailing perception that increased role dependency and decreased role autonomy may both lead to increases in team conflict. As many teams experience team projects, based on the data presented here, projects bring with them an inherent lack of role autonomy, and a necessity toward multiple team roles. The likelihood of conflict is therefore high when working in a virtual team, and a high propensity toward conflict may be aggravated due to the high potential for team members to lack exclusivity in their role, potentially leading to a greater level of interdependency in both task and responsibility.

With high levels of conflict may come high levels of stress and uncertainty, and perceptions of equity may suffer. Dependence upon perceptions of procedural equity during times of uncertainty, according to Hakonen and Lipponen, tends to be greater in teams with greater dispersion and low co-location contact (2008). Procedural equity, or a perceived fairness in the means by which decisions are made, is related to team members' perception of equitable process regarding decision making, and may be directly related to the development of trust and commitment, thus identification with the virtual team.

Issues of perceived equity may impact the level of conflict within a team, and, as Hinds and Mortensen note, may play a greater role in the virtual team setting, and is often understood to be more difficult to resolve than conflicts within collocated teams (2005). Collocated team members' development of trust is also significantly identified with perceptions regarding equitable decision making, and may affect team pride and self-respect, further affecting processes of vulnerability and self-disclosure.

Montoya-Weiss, Massey and Song, tell us that use of avoidance in a virtual team as a means of managing conflict may be detrimental to team effectiveness (2001). Avoidance, managers should always remember, is a very tempting way for team members to react to conflict when working in a virtual environment. As conflict arises, and employees seek to avoid further conflict, they may tend to make decisions that are detrimental to team success, and seek out greater isolation from the team . Healthy conflict resolution methods should be stressed.

Managers should seek ways to help team members to face conflicts and work together to find effective methods of calling out conflict and working through the cause. Competition among team members and other teams may be a good thing as teams may work toward a common approach to resolve the conflict when a majority of the team members are able to come to agreement (Montoya-Weiss et al., 2001). Collaboration is likely, as one may expect, to be the stronger of the conflict resolution practices. Collaboration, though, requires thoughtfulness and preparation of tools and methods by management as the project is planned and executed (Montoya-Weiss et al., 2001).

8

How Different Roles are Affected by Virtuality

Unusual but true, and previously unconsidered, is the recent finding that the role an employee may play on a project team, or perhaps in the office environment, may have an effect on the way in which trust is perceived (Wise, 2012). Roles that were analyzed in this study include project manager, engineering, analyst, developer, and tester.[1]

A survey was conducted with more than 500 persons that have recently participated in a team project. Participants were asked to identify the amount of time they normally spend in a typical day acting in each of the roles of tester, developer, analyst, engineer, and project manager (Wise, 2012). For each of the roles, a short definition was provided as follows:

- Tester was identified as a project team member that performs the duties of designing, writing, or implementing tests of software or systems.

- Developer was identified as a project team member that performs the duties of designing, writing code, programming, unit testing, or code reviews.

- Analyst was identified as a project team member that performs the duties of modeling, requirement elicitation, needs analysis, or system and hardware design.

---

1   "The purpose of this study was to find evidence that employees in the project role of systems and software tester may experience less effect on their trust of team members in other project roles when working in a virtual team setting" (Wise, 2012).

- Engineer was identified as a project team member that performs the duties of need analysis, software, system, and architecture design, or architect.

- Project manager was identified as a project team member that performs the duties of resourcing, scheduling, planning, tracking, or coordination.

As you can easily see, many of these roles may overlap, which as we have discussed several times already, may add to conflict and ambiguity in roles and responsibilities. While the overlap may be inherent in the work, these issues may be aggravated in a virtual office environment. So what, you might reasonably be asking yourself, is the point of this discussion? Simply that, in this research study, evidence was found that the role a project participant may play on the project team may have a statistically significant impact on how trust in each of the trust bases is perceived or prescribed (Wise, 2012).

To provide some context to this discussion, the study was designed to seek evidence as to whether team members working in the role of tester may experience a level of trust that is different than that experienced by team members working in other roles. Each of the employee roles were analyzed in each of the three bases of trust. Trust was evaluated in the three bases described by Sarker et al. as personality based trust, cognitive based trust, and institutional based trust to determine, when the response for each of the roles was compared to the response of the tester, if there is a statistically significant difference in the way that trust is perceived. The answer is yes. In many cases there is a difference in response by participants whose main role was that of tester from those who indicated other roles than tester as their primary role providing ample evidence that the role an employee plays may indicate that a difference in the perceived trust exists.

As a manager approaches the task of determining how to work with team members in a virtual setting, consideration needs to be given to understanding the role that the person plays on the team. Many team members working on an information technology (IT) project team will play multiple roles in the project. Due to the role ambiguity that this may cause, the role often changes based upon the work that must be accomplished at a given time, and therefore the approach the manager brings to the table should be tailored to the situation. In this study, most every respondent to the question regarding what role they recently played on a project reported playing more than one role (Wise, 2012). See Figure 8.1.

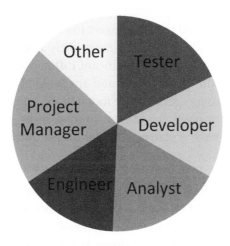

Figure 8.1   Distribution of project roles

The data in Figure 8.1 reflects the reality that the survey responses describe. In a given day project team member roles may change in order to fulfill the goals of that day, and the work they need to accomplish to fulfill those goals. Participation was fairly well distributed among the roles in the study. Only 12 percent of the respondents in this study indicated that they participated in their latest project by working in only one role (Wise, 2012). Where participants in the study did indicate an exclusive project role, the distribution of the roles was very different. Most study participants that identified an exclusive role reported that role as project manager (Wise, 2012) (see Figure 8.2).

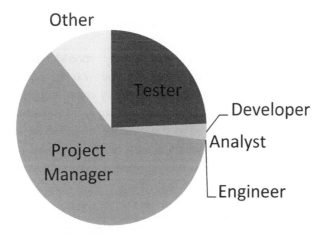

Figure 8.2   Distribution of project roles when an exclusive role is identified

This same study showed that the project manager, which is the most often identified role in the study, also is the most likely role in which participants were found to have a statistically significant difference in trust. A statistical difference means that the differences found are not likely to be due to random chance, but rather by some special cause found in the sample of the population. Even in the case of personality based trust, in some cases, employees in specific roles were found to have a statistically significant difference in the level of perceived trust.

Other roles were found to be very different in their level of trust based on the survey as well. That of analyst and engineer often responded with results that are significantly different than the responses returned by other roles as well. In each of these cases the responses at times were independent of the setting, however when the elements of virtuality are injected into the scenario, the results are dramatically increased in regard to differences in trust based on the role of the respondent.

Implications for Managers

When we look closely at the findings of the research described briefly above, the findings in this study interestingly appear to reflect a difference in the effect of a virtual team setting based on the role that an employee plays in a project team. It appears that some employee roles such as project manager may be affected by the virtual team setting in different ways to other employee roles.[2] Realizing that the role in which an employee participates may have an effect on the way in which the team member perceives trust, fellow team members can play an important part in team development and formation.

As managers establish project teams, they should take into account the geographic setting, and plan ahead for the way in which teams will communicate, and the information needs that may come with each of the roles represented on the team. This includes the degree to which teams may rely upon electronic communications, as well as the cadence, or schedule, and rhythm of communication. Also equally important may be the potential roles members play, and the setting in which the team is proposed. We as managers cannot lose

---

2   In the case of this study, statistical significance was measured at .05 indicating that 5 percent of the time the findings would be expected to be found in the sample by random chance. This may provide confidence that 95 percent of the time the findings indicate something significant in the sample.

sight of the possibility that these same differences in team member needs and support may often be present independent of a virtual setting (Wise, 2012).

Remember that virtuality is not dependent upon the setting, but may often be dependent on how people work rather than on where people work. The study referenced here found that, in some cases, the differences found in trust may appear independent of all elements of virtuality in regard to institutional based trust. Institutional based trust, if you remember the earlier discussions, is a reflection on whether the employee perceives equity in the application of rules, procedures, and practices of the organizations.

As we break the discussion down a little further and begin to look at the different roles we can see that the project manager does account for about one-third of all the significant responses received in the study(see Figure 8.3). The project manager is a leadership role and requires a great deal of vision into the project including the activities and progress of other team members. As we have noted several times, the act of leadership in a project may be distributed among several participants, and may change from person to person dependent upon the state of the project and current activities.

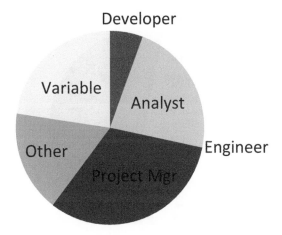

Figure 8.3   **Distribution of significant responses by percentage in the study**

Distributed leadership, according to Spillane (2005), is more about the act of leadership, or otherwise said to be the practice of leadership, rather than about the person who is leading. Leadership is about the way and means by which people interact (Spillane, 2005). Interaction between the people and the situations in which people find themselves is the essence of leadership in a distributed leadership scenario (Spillane, 2005). Spillane describes the action and interactions of leadership in a distributed model as a form of reciprocation in dependency (2005). Reciprocation is necessary to support trust relationships, and as we know from earlier discussions we have had, is heavily dependent upon the ability to discover information in a given situation.

This difference in trust also appears to be strong in the roles of analyst, as well as those team members playing a contingent role that is very highly likely to change based on the project situation. The *variable* tag is an indication that no one specific role was indicated more often than another for these respondents. It is interesting to note that the role of analyst shares many of the same trust issues in a virtual setting with the role of project manager. This may be due to the dual role most analysts play on a project. We see this duality in Figure 8.4. One perspective on this situation is that an analyst may often share the role of project manager depending upon the circumstance of the project.

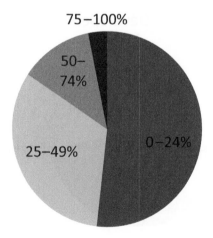

Figure 8.4   Distribution of percentage of time spent in the analyst role

Eighty eight percent of respondents in the study indicated that they spent time on the project in the role of project manager, which supports the activities of distributed leadership described by Spillane (2005). The United States Department of Labor report describes the analyst role in leadership terms such as diagnosing problems and recommending solutions as well as working to determine whether programs have achieved goals (2011).

Leadership roles at all levels of the organization often struggle as they seek to discover information that helps in decision making and orientation to the progress and cadence of the project. Distributed leadership, Spillane explains, is often dependent upon the relationships that each of the participants in the leadership team build as they work through problems in goal attainment (2005). While personality based trust does not appear to be particularly troublesome, a lack of ability to establish rapport may hamper necessary team activity.

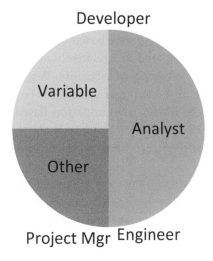

Figure 8.5   **Distribution of significant responses by percentage in the study for personality based trust**

IMPLICATIONS FOR PERSONALITY BASED TRUST

Interestingly, personality based trust, which researchers and psychologist tell us developed in the formative years of the employee's development, does not appear to be greatly affected by the elements of virtual based team activities. This may be due to the enduring nature of early development expectations of trustworthiness, and the expectation implied by co-worker's words or behaviors of fulfillment of the trust covenant. Personality based trust does not seem to be strongly affected based on the role that an employee plays with a virtual project team.

The elements of virtuality such as geographic proximity, communication, and culture do not seem to have an effect on the expectation of fulfillment an employee holds when evaluated based on the role an employee plays on the team. Perhaps this is due to an employee's self-perceived trustworthiness of others, or in part may be related to the person's own personality traits. As managers, we must always be aware that each employee brings with them to the work place their own experiences and personality.

IMPLICATIONS FOR COGNITIVE BASED TRUST

Cognitive based trust, as we discussed in earlier chapters, is a decision to trust based upon current and past experience and the expectations for similar future experiences. When we look at cognitive based trust we have a different picture of the trust response as compared to a relatively mild response to personality based trust regarding the elements of virtuality. Persons working in a virtual like setting may tend to have a stronger reliance on cognitive based trust, and this reaction does appear likely to be impacted based on specific employee roles (Kanawattanachai & Yoo, 2002). Roles in leadership capacity seem to be those with the biggest difference in the level of perceived trust.

Remember that the reactions we are talking about are in relation to the specific role of a software or systems tester. As shown in Figure 8.6, those team members that participate in the project in a leadership capacity such as project manager have a significant difference in the level of trust.

In this case, the perceived level of trust is significantly lower than that of the respondents in the role of tester. On average, respondents in the role of tester agreed with the cognitive based trust statements 85 percent of the time. In contrast, respondents that identified their role most often as a project manager agreed with the cognitive based statements only 70 percent of the time. Analysts' trust faired even more poorly with an average percent of agreement of 68 percent. This is a drop in the average number of positive agreed statements in perceived trust of 11 percent for project managers and a whopping 31 percent for the analysts.

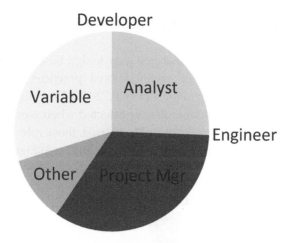

Figure 8.6   Distribution of significant responses by percentage in the study for cognitive based trust

Cognitive based trust questioning in the study were focused on the category of unit grouping. Unit grouping is the process in which team members form interdependent relationships and establish a self-identity within the group, or come to see themselves as members (Sarker et al., 2003). The formation process, as we continue to look closely at this research study, appears to be hindered in the case of some employee roles due to the challenges team members may face when separated from other team members, and when the perceived degree of cultural homogeneity is low. Managers should take note that levels of perceived trust lower for those team members other than that of tester, were reported more often when team members reported spending less time collocated with team members, and when levels of perceived homogeneity dropped. Cognitive based trust, however, improved as the use of electronically mediated communication increased.

This may be an indication that the introduction of electronically mediated communication for those employees in leadership roles actually improves the ability to discover information. As we know, the ability to discover information regarding co-workers provides for greater sharing and self-disclosure, and improves team formation. It may then be very helpful in forming successful project teams for a manager to pay very close attention to the availability of electronically mediated communications for their teams.

IMPLICATIONS FOR INSTITUTIONAL BASED TRUST

Institutional based trust is trust that one may assign based on the perception of equity in the application of organizational practices, procedures, and policies in the work environment (Sarker et al., 2003). Trust in this case may, as prescribed by employees in some roles, be affected when working in a virtual setting, and as is the case with cognitive based trust, those roles with leadership responsibility appear to be most affected (Wise, 2012) (see Figure 8.7).

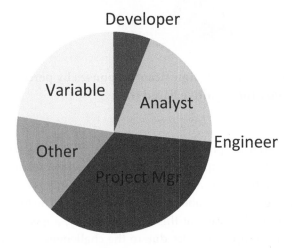

Figure 8.7 **Distribution of significant responses by percentage in the study for institutional based trust**

In the corporate setting, it may be that institutional based trust is in part dependent upon the propensity of a team member to trust, as well as the degree to which expectations for equitable application of corporate norms are met. The ability for employees to confirm equity in the way in which guidelines are applied and reinforced regarding what is acceptable behavior in project practices may not be as heavily dependent on the ability to discover information, which may, in practice, be somewhat hindered due to the degree of collocation and electronic communications.

Virtual teams, as we now understand, can be challenged in their attempt to discover knowledge of their team members in many areas such as personal behaviors and experiences, expectations, project milestones, as well as work output to name only a few examples. This includes project relevant information as well as information that supports a social relationship in the work setting. Due to geographical separation and the degree to which electronically mediated communication and technologies are used, the challenge in gathering information to support a needed level of institutional based trust are often aggravated.

Perhaps in frustration and the need to get work done, teams in a virtual setting will often tend to rely on cognitive trust due to a common lack of face to face interaction or organizational knowledge. This may be cause for a reliance on cognitive trust, however the driver behind a reliance on cognitive based trust should drive managers to seek better ways in resolving the problems in identifying institutional equity as the elements of virtuality are adopted. Teams should be given the tools and vision that supports the perception of equity in the ways in which business policies are applied across the organization.

As we look closely at the affect the elements of virtuality have on institutional based trust, we note that the frequency of statistically significant results dropped with the introduction of the variable geographic proximity. Purely speculation, however I find it very interesting that as team members are separated they appear to be less concerned with the equitable application of corporate policy. Perhaps what cannot be perceived becomes less important when we are talking about fairness? What may be more damaging is the possibility that this change may signal a level of frustration, or perhaps it is simply a case of out-of-sight-out-of-mind?

In either case managers should be concerned. Institutional based trust is a significant indicator that employees are content and satisfied with their relationship with the management and policies of the organization. This relationship can have an impact on retention and dedication to the success of the overall organization.

With the introduction of electronically mediated communication the difference in perceived trust widens and the incidence of statistically different responses increases. It appears, and again this is purely speculative, that as communication increases the concerns for equity are enhanced through communications that are somewhat devoid of human social queues. Or perhaps as information is made available, team members do not like what they are able to find? It may be that, with the cryptic nature of electronic communications or the discovery of more information, more questions are raised than answered?

This may be driven by the idea that information necessary to create a solid understanding of the state of affairs is not readily available. Here is a brief example, one close to my heart as I have often been involved in these sorts of corporate practices. One of the practices of many organizations is to survey employees about their feelings regarding corporate polices, processes, and practices. Most management view this information very seriously and do change or adjust the way in which things get done based on the outcome of these surveys.

Unfortunately, sharing the information about the changes that result as an outcome from the surveys is not viewed as seriously as the original feedback that employees provide. Management may send out an email, and may address the information in management meetings, but getting that information to the folks on the project teams is a little more difficult. Team members are left to feel as though nothing ever changes because the information does not reach them, and the systems of information knowledge sharing tend to be somewhat ineffective.

Remember that those that seem to have the greater challenges in institutional based trust are those in leadership type roles. These are the people with influence and a strong voice in the team. This seems to lead us to speculate that, as information is discovered regarding ways in which institutional practices are applied across the organization, more questions regarding fairness are brought to mind. Those questions that seem to raise the greatest concern for the project manager are around timeliness, effort, and dependability regarding deliverables.

As for cultural concerns with institutional based trust, differences in culture, whenperceived as a degree to which team members are alike, does not seem to be a major issue regarding organizational equity. When the element of culture is introduced independent of other elements of virtuality, the number of significant results dropped. This leads to the assumption that the degree to which ethnicity or corporate culture contributes to concerns regarding trust is not necessarily a hot issue (see Table 8.1).

Table 8.1   Institutional based trust frequency data for statistically significant results

Institutional Based Trust		
Geographic Proximity	Communication	Culture
9	19	6

In the current work environment, where it is often a common practice to communicate primarily by electronically mediated means, managers need to be highly sensitive to issues of perceived equity in institutional practices and norms. The issue of awareness may even be described as critical when it comes to those employees working in roles of influence or control. Project managers and analysts have the ability to influence a positive or negative outcome when it comes to project success, and must have confidence in the institutional practices that govern their project lifecycle.

If managers are not on top of the issues around institutional based trust, the use of electronically mediated communication may increase the separation between perceived groups in the work place. This may be due to attempts to evaluate the non-verbal signals in communications and trust building opportunities as they relate to social norms, and since there is some evidence that trust may be cumulative in nature, the absence of social norms in mediated communications may serve to increase boundaries between work teams. This may be a confounding variable since, Kirkman and Mathieu note (2005), collocation may not lead to anticipated collocated behaviors, but rather, a team may choose to behave as a virtual team regarding communication patterns.

IMPLICATIONS FOR MANAGEMENT PRACTICE

Managers should consider, as they prepare their plan for the next project, a seven factor plan designed to establish a management strategy for team success as a virtual team framework to understand the relationships between teams and trust (Duarte & Snyder, 2006). The seven factors of human resource policy are training, processes, collaboration and communication mediation, culture, leadership support, and leadership and membership competency. When evaluated with virtual teams in mind, these seven factors may guide managers in understanding a global team model (Duarte & Snyder, 2006). Effective practices in each of these areas may support the needs of those project team roles that appear to be most vulnerable in regard to trust in a virtual team setting.

In a virtual team setting managers may need to attend closely to several necessitating factors of virtual team success, such as facilitating face to face gatherings on a defined schedule, a shortened planning horizon of activities with lesser complexity and intensity, and shared leadership (Lipnack & Stamps, 1997; Maznevski & Chudoba, 2000). The practices suggested in the framework may provide a useful framework when considering the needs of a team member performing the role of a project manager or analyst.

Because virtual team members are very dependent on written communication, they cannot see the nuance of facial expressions and body posture, and may be expected to merely take the written rules at face value. They are often missing the stories and corporate lore that is shared in the meeting room and around the coffee pot. We may, as managers, make an assumption that rules may be reinforced when human resource emails or letters are received regarding institutional changes or updates, promotion notices, employee role changes, and publish notices of organizational changes in structure. Building an integrated plan based on the seven factors identified by Duarte and Snyder may help establish awareness in the management ranks, and set a strong base within the organization to support virtual work practices.

9

Understanding Quality Assurance: The Basic Mechanics to Support Information Discovery

Over the last few years it often appears that many management professionals are confused as to the practices and contributions of quality assurance (QA) and quality control (QC), or worse yet are not aware that they are different. This may really be just a problem of scope, and at times a problem with the basics and history of quality management. So, let's back up a moment and do a little background check regarding the relationship between quality, teams, and trust. It is important to have an understanding of the different roles that QC and QA play in order to understand the real impact that QA may play in building a trust relationship. To make the discussion simpler, we will spend some time talking about software development as a project. I will use this approach since software is a common key component in almost every device we use.

A Little Quality Control Background

When we talk about QC, what does this mean? Generally, when we talk about QC, we are talking about random sampling, or commonly known as testing. Sampling, as Deming describes the practice, starts with the assumption that a sample represents the population, and provides an unbiased estimator in the analysis process (1976). The sample is expected to represent the degree to which the entire product may satisfy the customer. In social sampling the population is represented by a selection of people whose opinion, activities, or condition may represent those of us not lucky enough to have been chosen as part of the sample.

An effective sampling process begins with the assumption that the entire population under study has an equal opportunity to be chosen in the sampling strategy. Randomness, when we talk about a random sample, is really an expression of the likelihood that the process used to select the sample from the population provides an equal opportunity for all members of the population to be chosen as part of the sample. The process of sampling is important in an industrial setting to avoid unnecessary interruption to operations, as well as in social, knowledge, food, and software systems production, and in most every product, delivery, or service industry. Sampling provides a statistically significant probability that the sample represents the whole of the population, and offers the best opportunity to control the output of the operation.

In production of most any product the sample represents the product produced in a specific configuration. Or, as in the case of software development, the sample is a specified suite of tests that are run against an instance of the application, or build, in a specified configuration. Sampling in software requires extensive planning to ensure the sample set includes tests that may be considered representative of the population of transactions that the software application, component, or system may experience in operation. The final product release may be considered the population in its entirety, and the test configuration may be considered the sample of this product. This includes, just as in industrial production, tests to verify requirements and specifications are met, as well as functional, reliability, performance, security, load, and destructive testing to ensure failures are controlled and managed gracefully.

So, QC, in a discussion of software, is considered to be the test. Our discussion of QC, or testing, then implies a basic standard operating procedure for development of a test. A test, or QC, begins with a question for which a series of events are then defined to answer. The environment is established to support the type of testing, and a correct response for each of the tests is predetermined. It is important to note that without a predetermined correct answer, a test has not occurred. Without the correct answer, the activity of testing is no more than a query.

A BIT OF QUALITY CONTROL BACKGROUND

How QC is accomplished is often tailored to the way in which work is planned and executed within the engineering process, and therefore is very highly dependent on how the process of accomplishing the work is managed. This process is called the lifecycle in software development circles. Software and systems development teams have taken on many lifecycle methods over the years ranging from traditional waterfall, iterative cycles, Rapid Application Development (RAD), and agile to name a few of what may be considered the most oft mentioned and widely understood methods.

Each of the methods requires some changes in test tools· and team configurations, as well as differences in member skills in order to achieve success in testing. The tester role has remained a key resource in project success throughout the changes, yet has taken on differences in participation and positioning within the organization. Additionally, testing methods may become more dependent on software automation in methods such as agile product development.

In order to fully understand QC, we have to consider the lifecycle within which the product is engineered. Waterfall software development is named for the cascading flow of the product as it moves through the lifecycle phases of requirement, design, development, test, and deployment, in a series of completion decision points, known as gates. The product being developed becomes increasingly well defined in each phase of the lifecycle and in complexity and degree of completion with each gate (see Figure 9.1).

In the waterfall lifecycle the tester is generally positioned as a part of a separate organization from the development team, and is considered to provide an impartial assessment of the quality of the product, due to their separation from the engineering function in order to perform a non-partial and independent verification of requirement conformance. This places the tester's role as the last decision point activity prior to product delivery. In a waterfall lifecycle, QC is often faced with a compressed timeline and reduced resources as a result of earlier project challenges which often contribute to tester versus developer angst (Perry & Rice, 1997).

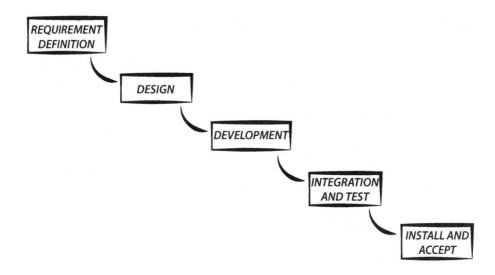

Figure 9.1   The typical waterfall lifecycle

The single hardest part of building a software system, according to many researchers, is often considered to be the level of complexity of the applications and network design. As systems complexity has increased so too has the need to improve the ability of the organization to describe the design and user required attributes of the system.[1] Engineering and development teams, in response to the level of complexity, have continuously sought ways to increase the cycles in their methodology[2] allowing for improved requirement discovery.

This method is commonly referred to as iterative development which uses repeated process loops through the lifecycle. In each of the loops through the process the understanding of the product design increases, and therefore the completion and specificity of product requirements are expressed in more detail with each cycle. This repetitive discovery process takes advantage of the increased knowledge of the software by the designers and developers

---

1   The design and user required attributes of the system or application are commonly referred to as business requirements and technical requirements. Requirements define the functions that a system or application need to be able to perform in order to be considered fully useful. Requirements are often referred to as "described" or "expressed" by IT practitioners, and referred to as "discovered" as more information about the application or system is understood during the design and development phases of the project.

2   A methodology is a set of related methods, or "tools" that may be used to accomplish a piece of work.

over repeated engineering cycles, and aids in understanding the business and technical needs.

With each iteration, testing is planned, executed, and defects reported and repaired, thus providing immediate feedback on software development progress. Using an iterative process allows the development and design teams to take into account the experience and increased understanding gleaned from the previous cycle. Iterations are similar to the waterfall method, yet not quite the same as a single pass through a waterfall lifecycle since each successive iteration addresses only a small portion of the entire product design. In theory iterative testing is deeper and more comprehensive when requirements are tested in smaller, testable deliveries. The testing, or sampling, remains a neutral process in the iterative lifecycle.

The testing process is generally accepted to be dedicated to the verification that the system or application conforms to the business and technical requirements. However, as development has become more rapid in order to keep up with the competition and the pace of changes in technology, and application complexity continues to increase, the sampling process has moved positionally toward the development team.

RAD has the team of developers and testers working together as one cohesive engineering unit. The developer, product end user, and analyst define and build the software system in a sequence of short, time bounded, efforts in an interview format to identify requirements. The applications and systems developers work in concurrent development teams and employ engineering techniques such as design of prototypes to guide the design of the product to shorten the process.

This lifecycle approach to software development draws expression from a business perspective to ensure all requirements are described and fulfilled, and works from an information technology (IT) perspective to ensure components and controls are in place to identify and express any missing requirements (Reilly, 1995). Users and customers will sometimes take on some testing roles. Unit testing, which is normally a developer test of the smallest testable piece of code, is at times a customer role to allow developers to continue development work. Formalized testing will often overlap the development process throughout the project.

It is a common practice in RAD development to maintain the tester role as an independent entity. As RAD methods grew in acceptance and matured, methods using trained project planning facilitators evolved into agile development and took hold. This shifted the development process to short two week development efforts, called *sprints*, with well defined methods and goals defined for each sprint.

Agile methods changed the look and feel of testing making the tester an integral and essential member of the development team (Talby, Keren, Hazzan & Dubinsky, 2006). According to Talby et al., testing and development are dependent functions, fully integrated, molding the traditional role of tester and developer together as a writing team that includes the engineer, analyst, and project manager together with the product owner (2006). The agile tester no longer plays the traditional *police* role, and morphs to that of helper to assist the team in documenting requirements through validation of the software capability (Crispin & Gregory, 2009).

A Little Quality Assurance Background

When we talk about QA, and not QC, we are talking about all of the processes and roles that encompass the activities that take place to be sure that when we produce a product, we actually produce the product in the way in which it was intended to be produced. QA is the practice of assuring that the actual activities of creating the product are conducted in the manner in which they were intended to be conducted, and to the design and specification as it is defined. Key elements of every QA program are the identification, generation, categorization, analysis, and presentation of relevant and useful information designed to provide a projection of the probable outcome of the project, and the state of the process.

QA ensures that the processes and activities of the team, group, and product delivery cycles are visible and understandable, and that key activities are carried out according to management expectations. Through this process managers are also able to assess the capability of their processes to deliver a product within the constraints, or desired specifications. In choosing to use a fully functional assurance program, managers have then chosen to employ a set of tools that include measurement, analysis, and dissemination of information designed to bring transparency and openness to the targeted programs.

Merriam-Webster notes that QA is a term made common around 1973 to describe a program of systematic monitoring and evaluation of work processes (2010). While there is a common set of QA practices that most organizations recognize as assurance related, there is no truly recognized standardization regarding how to define QA as a program. QA programs often include the standardization of procedures and measures, with specific processes for evaluation and analysis with the intent of driving improvement and ensuring quality standards are met.

The common practices of QA that transformed manufacturing processes into a highly effective and repeatable set of production methods have been largely absent in the software development industry. Software Quality Assurance, similar to QA in any industry, is designed to address the need for assurance that the lifecycle, as defined by policy, is effectively implemented to avoid the cost of finding defects during testing or deployment. In addition to the role described above, many QA practices have evolved to one of statistical analysis toward the goal of reducing process variation to provide stability and increased process capability. The challenge for QA practitioners is to add value to the process while avoiding intrusion upon those working through the daily activities of getting the work done.

SOME HISTORY OF QUALITY ASSURANCE

In 1995 Carroll described software development as a creative endeavor reliant on the imaginative ability of the software developer, and proposed that the application of Total Quality Management (TQM) be applied to integrate the research and design practices with development and quality. TQM was the new wave of ingenuity in the early 1990s with new quality theory breaking across the bows of every Fortune 500 company. IT and software development practices were not exempt or given exception, with many IT companies attempting to embrace the TQM momentum.

TQM is participative by design, and is built upon the basis that quality is the responsibility of everyone, thus focused on the practice of self-management and autonomy, as employees go about the planning and execution of their daily work (Brooke, 1994). TQM has been described by some as holistic in practice and theory, and designed to engage workers in a multidisciplinary approach to defining and implementing quality. TQM became the new way in which companies moved mountains.

According to Brooke, TQM practitioners were working to draw into the new wave of quality, the production of IT services and products from the traditional practices of QA and QC toward the acceptance that quality is everyone's within everyone's bailiwick (1994). Executives responsible for the production of software began to accept the treatise that acceptance of fitness for purpose, or that if a product is capable of performing the actions for which it is designed, then that is the determination of a quality product, and not the absence of defects (Brooke, 1994). With the diffusion of quality responsibility across multiple work groups and individuals, information design, collection, and dissemination became a key function of QA.

With the need for greater information collection, and analysis came the realization that information regarding how work gets done is an essential component, and rediscovered the idea that all work functions, for better or worse, within a set of constraints called a process. Around this time managers of quality rediscovered that process, and the management and application of process, was the key to inexpensive and effective quality. McManus and Wood-Harper (2007) noted Juran's[3] exclamation of quality in 1988 in restating that the vast majority of problems, even up to 90 percent of all problems in companies, are systemic in nature.

That is, problems are driven by poor process, or poor process management. With this understanding managers began to realize that process problems are not necessarily caused or influenced by staff practices. This became known as the eighty–twenty rule. The eighty–twenty rule is still the rule of the day for quality practitioners, and is interpreted as 80 percent of the problems can be eliminated by finding and eliminating the top 20 percent of the key process problems.

Seventy-five percent of software projects fail, according a Standish Group study, due to causes of failure such as coming in late on schedules, over budget, or due to poor quality (Cramm, 2001). Cramm goes on to say that many failures are combined in a cumulative fashion, and become magnified by a domino effect with past failures due to potentially all of the above (2001). The failure rate of larger projects such as those projects costing more than $10 million are considered by some to be approaching 100 percent (Cramm, 2001).

---

3   Juran, in discussing the mathematical formula created by Vilfredo Pareto in 1906 describing the unequal distribution of wealth in Italy at the time, coined the phrase the "Pareto Principle" to describe the expected distribution of cause and effect. The "vital few" causes, Juran observed, are often responsible for the "vital many" problematic effects.

Improvement on the failure rate often requires the use of proven project and quality methods designed to increase predictability of schedule and product quality, and ultimately the opportunity for success. In an attempt to find the means to manage processes managers began to reach out to users hoping to find ways to better understand their experiences with software products. According to Peskin and Hart (1996), a key driver of software project difficulty is a result of not taking the time to understand the user's needs, and perhaps the ways in which the user intends to use and apply the capability of the product to solve their problems.

QA needs an early start on every project to deliver the greatest impact. In the beginning, or concept phase, at the start of each project is the proper starting point at which QA must be engaged to ensure the customer's needs are effectively captured. Starting early is the best way to avoid problems that will undoubtedly create delays if not found in the earliest days of the project. Any delay in capturing the customer's demands may add to the project duration. QA professionals became engaged in providing assurance that the user information and experience were effectively captured and shared with the product designer. Early QA involvement ensures that project quality expectations are well defined, and QA programs are well prepared to meet the information discovery and dissemination needs of the project team and organization. Once again the question of information discovery is key to answering the problems of quality.

QA is best focused on building effective and repeatable processes and information discovery. With this in mind, when quality professionals began to look at project failures they discovered that each project failure tended to display specific failure modes and mechanisms. Most of these failures, Kruchten notes, can be categorized in a combination of root causes such as ad hoc requirements, communications failures, failed architecture or extreme complexity, inconsistencies in requirement and design with the implementation, testing shortfalls, realized risk, and change management (2004).

In keeping with the TQM theme of the time, Kruchten notes that "Visualizing, specifying, constructing, and documenting a software intensive system demand that the system be viewed from a number of perspectives" (2004, p. 9). The many perspectives of the developers, users, testers, and analysts, as well as others continuing into all participants and stakeholders, each address the project in specific and unique ways that require transparency and participation in, and understanding of, the system architecture (Kruchten, 2004). Again with the need focused on transparency and information discovery.

The Role of Quality in Information Systems/Information Technology Delivery in Modern Strategy

In recent years quality in information systems (IS) and IT delivery processes have re-established focus on three distinct areas of concern described as risk management, process quality, and QC, or testing. These areas are complimentary practices, and are highly dependent upon the ability to discover and disseminate information to project team members. Risk management is completely reliant upon the ability for team members to find and understand comprehensive information in resources, schedules, and dependencies. Process quality requires information regarding process effectiveness, efficiency, productivity, and quality. QC requires extensive information such as requirements, specifications, defect discovery and repair, as well as schedule, priority, resources, and other areas of project planning.

The subject of risk management in IS and IT delivery is as broad as it is deep, and therefore needs to be narrowed, as it pertains to our discussion. In this context we will focus on the portion of risk analysis as it relates to determining the satisfaction of requirements in critical features of a system or application, and the management of that risk through stakeholder involvement and targeted verification and validation of those features. Risk in this context is the assessment of the remaining risk in the product at the point of delivery or provisioning to the production systems or customer site. This requires the ability of the project team to identify the degree to which risk threatens the ability of the product to fulfill the user expectations, or exposes the business to failure. QA practices, as described by McManus (2004), provide a brief description of this reasoning.

McManus identifies stakeholder involvement in the strategic planning and assessment processes for development of system and application requirements as absolutely essential in the reduction of delivery risk (2004). Project management at its very basic level is the development and coordination of the plan including scope, scheduling, resourcing, and risk identification and mitigation. Research would suggest, McManus continues, that one common characteristic of unsuccessful software projects is that planning is conducted in some degree in secret and without stakeholder contributions (2004). As a result of lack of stakeholder involvement, increased risk is often inevitably realized as customer rejection of the end product, and thus a failed project.

Through the use of effective QA practices such as those found in TQM programs, stakeholder involvement in the development of project information planning and sharing is sought out and incorporated early in the planning stages. This can be done through relatively simple practices such as surveying or hosting focus groups for internal as well as external customers and involving project team members from all involved groups in the planning stages. TQM methods involving project planning become ever more important as teams become more geographically dispersed and virtual in their work practices.

In researching global project management, Anantatmula and Thomas (2010), address risk in the light of the relatively new movement to move many roles in IS and IT projects offshore. Leadership and establishment of trust, according to Anantatmula and Thomas (2010), are absolutely essential in the reduction of risk due to several aspects of working offshore that are not present in the relatively homogenous environment of in house, in country, project teams. "It is critical, in the context of global projects and geographically dispersed project teams, to integrate information technology [IT] tools and manage cultural differences in dealing with project risk and complexity with a focus on improving efficiency, effectiveness, and innovation" (Anantatmula & Thomas, 2010, p. 1).

Cultural differences, Anantatmula and Thomas suggest, may at times manifest themselves in the areas of increased project risk if trust between team members is stressed by cultural differences aggravated by a lack of face to face communication. Global projects, simply by the nature of such projects employ more people that are separated by space and across time zones, inherently carry new problems in communication, and difficulty in collaboration. Overcoming the lack of information availability caused by the separation of team members requires a robust QA program of process quality.

When focusing on process quality, the discussion turns to metrics, lifecycle management and continuous improvement practices. For improvement of product quality processes, test management and test practices are often the main focus of management concern since these practices have a direct and visible impact on the product delivery lifecycle. Process quality, however, is the process of providing assurance that the process is well understood, tailored, and applied to the specific project, and capable of performing to expectations, then controlled through documented expectations and review.

Statistical Modeling of the process expectations and performance as a means of project control is increasingly used in the improvement of IS and IT project strategies. In 1999 quality management Researchers began to once again advocate for the use of the same statistical process controls and continuous improvement methods advocated by Shewhart and Deming.

IT development professionals have long aggravated quality practitioners in the ability to use metrics and statistical methods. At times it seems as though the practice of measuring software development practices and efforts impinges upon the creativity and artistic expression of those team members attempting to practice software coding. Measures such as lines of code per hour, day, or week have been met with resistance, and exclamations of dismay. The argument for the most part is due to a perceived inability to overcome the vast differences in complexity and challenges in defining what is considered "development work."

While this may not seem important, the ability to measure and share information regarding progress in any professional practice is essential in forming and supporting team trust. One way of looking at the predicament that the inability to effectively measure software development can create is through example. I will never forget the time I was called on the carpet for being two weeks late on testing a critical new product release. As the director for testing in a major financial institution, my department was responsible for test planning, scheduling, and execution for all new financial instruments for electronic trading in the exchange. We were waiting patiently for the first product build as the next project review meeting approached.

When it came time for the development team to report on their progress I was expecting the director of software development to report on his progress in writing the software. To my dismay, the first report the director made was that the test team was now officially two weeks behind in testing. I leaned over immediately to my test manager and asked, "Did you receive the code?" I was a bit upset at the thought that my test team was two weeks behind when we were only two weeks into the test schedule.

The test manager looked shocked, and protested with great indignation that her team could not possibly fall two weeks behind when only two weeks into the schedule. It then dawned on us both that what the software development director was reporting would be more accurately described by saying that the software development team had not delivered the code, and they were now two weeks late. The real problem was that their's was not a QA function collecting data and disseminating reports on the project progress. Everyone was left to find their own way of tracking progress, and the executive had no way of discovering and tracking the progress of their development organization.

Once we understood what the director was really saying I was able to redirect the discussion to better understand the circumstance of the project. I asked the director when the code had been delivered, and with a rather sheepish look on his face he readily admitted that the code was not yet delivered to the test team. This situation had gone a long way in damaging trust between the two teams. In order to remedy the problem of trust we established a way by which the development team progress could be better tracked and reported through automated data collection using configuration management tools as mentioned in earlier chapters.

Writers of software code often feel compelled to challenge quality professionals in addressing measures of work output due to the time spent in analysis and evaluation of coding methods and strategies that may skew measures of output. As Glen states it, determining what may be called *work* for software developers has always been a touchy subject since knowledge workers spend much of their time in thought (2003). Florac and Carleton, however, challenged this notion by asserting that statistical processes can and should be applied to all aspects of software development including the planning and analysis as well as estimation and design, coding and testing, as well as all the other areas such as peer and quality reviews (1999).

To accomplish the establishment of process control, Florac and Carleton (1999) describe the key management practices for process management as defining the process, measuring the process, controlling the process, and improving the process. In turn, these process controls can form the basis of a metric program designed for information sharing among team members.

As Deming explained in the past, all processes contain variation, and therefore can be managed using simple statistics for measuring the process variability and providing for predictable levels of quality in the result. Why is this important to us in this discussion? Because when we decide that information is to be collected, this helps guide the organization in understanding what information can be shared, and where the likely gaps in information may be hiding. Controlling a process through the use of statistical methods is designed to increase the predictability of process outputs by reducing the variation in the process, and sharing information regarding the predictability of a process aids planning.

There are two types of variation in every process. What is called random variation is found in all processes, and is simply the natural rhythm of a process as work is done. In every process equipment wears down, people get tired, and logistics falter. Even in a manufacturing environment where the process is designed to be repeatable producing the same product every time, there are small differences in measurements and material compounds as the product is processed. These changes occur by chance, and are normally not influenced by outside forces.

In the case of special cause variation, the process is affected by something outside of normal operation. This may be attempts at process change or a determined attempt to work around a difficult process, or in the case of IT perhaps the pressures to perform placed on the project team. Changes in the process, however, can be measured and tracked. Special cause variation in the process should be identified and removed through the use of problem solving techniques.

As quality practitioners are able to identify changes in the IT processes, they improve the capability of the process to produce good and effective products. The general goal of using statistical measurement to reduce variation in a process is to induce stability in the process to enable predictable results. Process stability as a measure of variability is based on the theory that randomly collected data, when collected from a process or population without the forces of special cause variation, will approximate a normal distribution curve.[4]

The normal distribution curve is a mathematical representation of normal process data distribution with known mathematical properties allowing for the determination of normalcy in the variability of the process measures. The normal distribution curve is the basis for the three different methodologies discussed by de Mast. There are three primary statistical methodologies, according to de Mast (2004), as proposed in IS and IT literature for the continuous improvement of IS and IT project outcomes. Taguchi's methods, Six Sigma, and the Shainin system are compared by de Mast beginning with Taguchi's methods in the 1980s.

Genichi Taguchi's methods were under development in the beginning stages of IS and IT production in the 1950s (de Mast, 2004). Genichi Taguchi proposed that loss due to poor product quality is not an event, but rather a

---

4   The normal distribution curve is characterized by a symmetrical bell-curve with a single peak at the mean, or average. This curve is used to illustrate the central tendency of data, or otherwise described as the tendency of randomly collected data to migrate toward the mean when the process for which the data is collected is not affected by outside causes.

gradual degrading of the quality tolerances. As process performance variation increases the ability of the process to meet quality tolerances and performance targets is diminished. Taguchi's methods practiced primarily in Japan until the 1980s were, and still are, at the forefront of statistical process control methodologies in process improvement (de Mast, 2004).

The Shainin system of statistical engineering begins with a problematic process output, and uses simplified statistical analysis to measure the variation of the process output to home in on the process problem. Using the Shainin system allows the engineer to use simple statistics in the analysis, and a methodology that avoids the complex language of statistics. Shainin methods are appreciated for their simplicity, de Mast explains (2004). The system's appeal is the ability for a person with an information technology background, or other technical discipline or background, and limited knowledge of statistics, to quickly grasp the methods and be able to apply them to their practice.

Based on the theory of normal variation within a process, Six Sigma is probably the most widely used of the three methodologies for process improvement. Characterized by a customer driven improvement approach, Six Sigma focuses on quantitative data collection and analysis with a priority on saving money. Six Sigma[5] uses nearly all of the statistical methodologies as tools and techniques available to the well trained and experienced practitioner in identifying the highest impact problem, and reducing the process variation as a means of controlling costs within the process.

Complimentary Goals and Roles

Each of the methods described in this brief view of QA and QC describes independent, but complimentary practices, that when taken together form the ability to generate, collect, analyze, and disseminate information that may support information sharing, and thus the development of trust. The roles of QA are complimentary, however not dependent. Each may exist separately, and provide effective process outcomes. However, QC without QA has a tendency to exist with a diminished ability to affect the quality capability of the organization.

---

5   Six Sigma takes its name from the statistical calculation of standard deviation. Standard deviation is a measure of the dispersion of difference from the mean for a given sample of data. The symbol for standard deviation is the Greek letter sigma (σ). In Six Sigma the expectation is that 99.999 percent of the data will fall within three standard deviations of either side of the mean.

PART III

Quality Assurance, Trust and the Virtual Team

Introduction

Our thoughts on quality and the role quality plays in projects and corporate identity have shifted and changed as the need to compete and carve out a niche in the market place moves with technology and societal norms. Additionally, many companies and managers have sought to find ways in which virtual teams may be supported through company policy and human resource practices. As practices in quality and corporate policy and awareness have changed over the years, so have the ability to support virtual project teams.

The next couple of chapters provide a brief yet useful overview of quality and quality practices in relation to teams and team projects as a way of providing context to the author's view of the project world. Understanding the different roles that quality has played over the last couple of decades, and the differences in quality practices may provide the reader with a foundation with which to approach the subject of quality and quality's relationship with trust and team performance.

Quality plays two primary roles in projects and processing within the office environment. The first role to address is that of a controlling force. Quality control (QC) is an intervening force designed to address the outcome of a process or project stage as a measure of output quality. Quality assurance (QA) is a mediating force designed to address the potential for a positive outcome of a process or project stage as a projection of the probable outcome. The following chapter provides some background on the shifting role and focus as the quality roles have matured and adapted over the years.

As the discussion in Part III progresses, the author shifts to an overview of factors that affect the potential success of a corporate strategy designed to gain the most from the virtual office environment. Factors such as human resource policy and training and development, as well as leadership policy, are briefly reviewed with the goal of providing management awareness as to the effect such strategies may have on the virtual team. How managers work, and the guidance and reinforcement of policy in these areas may have a strong and direct effect on trust in a virtual office due in part to the mitigating effect of the elements of virtuality.

The following chapters are intended to provide context to allow the reader to assess the ways in which an integrated management strategy may strengthen the virtual office. Team members may benefit from a conscious corporate effort to build a strategy that provides a comprehensive view of the corporate policy in relation to the way in which work gets done in the virtual world.

10

Organizational Strategies That Influence the Success of Virtual Team Management

In the year 2000 Wong and Burton noted few true virtual teams in the information technology (IT) work force. 2010, however, presents a very different picture with researchers identifying a need to address the heightened project risk due to the strong movement of IT projects to offshore and virtual teams (Anantatmula and Thomas, 2010). In working with projects in a global project management environment, leadership confidence and skill in the establishment of trust with and by leaders can be critical to reducing the risks caused by several aspects of working offshore that are not present in the relatively homogenous environment of in house, face to face, project teams. In order to address the risks of global project integration, effective communications and information discovery using application of technology and leadership has become a key element in project management (Anantatmula & Thomas, 2010).

In the 1990s smoke filled rooms gave way to the cordial, environmentally acceptable work place, and now, as technology in the 2000s take hold, work, according to Ahuja, is no longer a place people go, but rather something people do (2010). Technology allows for the lifting of barriers that restricted traditional jobs. Barriers such as geographic boundaries, cultural norms, and organizational practices and restrictions regarding the timing and the locations of employee contributions no longer limit how work gets done. As we explore the relationship between technology, motivation, and teamwork among software and systems developers, engineers, and testers and other quality professionals, an understanding of the team dynamics of motivation, communications, and belonging are highly relevant.

To help us understand the needs of the new team dynamics of a twenty first century project team, Duarte and Snyder (2006) provide a seven factor framework with which to address the conversation regarding management strategy and potential team success in virtual teams. According to Duarte and Snyder the seven critical elements of a virtual team strategy are:

1. human resource policy;

2. training;

3. processes;

4. collaboration and communication mediation;

5. culture;

6. leadership support;

7. leadership and membership competency.

These seven elements, when plans are formulated with policies that accommodate the needs of virtual teams and virtual team leadership, may provide greater success in managing the global workforce. With the elements of an effective strategy built into the tactical project planning of the organization we can find support among complimentary studies that will help managers formulate a strategy that is customized to fit your particular project and environmental needs.

Human Resource Policy

The practice of quality in IT and information systems (IS) has evolved from one of an engineering focus in feature and requirement verification, to one requiring membership with and participation in engineering and development practitioner teams. While this shift may seem natural to some, it does come with some new and difficult challenges in human resource management practice. In the past, IT and IS employees were considered to be motivated by their passion for their craft. With a shift to virtual team based planning and execution of projects, the silent individual must now become a member of a larger execution model.

In an article a few years back, Faircloth and Hamm recorded their studies of the motivation to achieve high marks in school in a group of high school students noting that membership, or as we have described it as identification of oneself with the group, as a significant variable in defining the relationship between motivation and achievement (2005). Trust has always been an essential part of teambuilding, and yet we are really only learning how working in a small world environment may affect team dynamics. This new team model shrinks our view of the world and expands our view of the team thus impacting the essentiality of trust in team dynamics as teams begin to rely more heavily on geographically dispersed team membership in building effective communications.

Feeling at ease within the group and at home with the role we play is necessary to building membership, and may be in some part supported and driven by equitable management policy decisions. Ryan and Kossek express this concept as an essential part of the human resource plan by saying that, to the degree that management may implement policy that eases the work and home life stressors, then employees may move toward feelings of membership within the work group (2008). Additionally, to the degree that managers may equitably implement these policies, this may in part have an effect on the degree to which an employee may perceive themselves to be a member of the organization or team, and feed team performance through higher levels of trust and commitment (Park et al., 2005).

Performance, according to Whetten and Cameron (1995), is a quotient of ability and motivation, and ability the quotient of aptitude, training, and resources, while motivation the quotient of desire and commitment. These five human resource variables come together to make the backbone of every project team, with the capability of impacting the likelihood of the team's success or failure. Virtuality, according to Ahuja (2010), is the new norm as organizations diversify across geographic boundaries, therefore straining the corporate ability to maintain a focus on the motivational factors in Whetten and Cameron's performance equation.

Pre-dating Whetten and Cameron and establishing a foundation for the desire factor within Whetten and Cameron's equation, Maslow (1948) provides motivational theory in a graphical representation of unsatisfied needs in a hierarchical pyramid based on the human desire to satisfy those needs of the most basic level prior to graduation up the hierarchy. Physiological and safety needs, according to Maslow, must be satisfied before the employee is driven toward satisfaction of needs at the social and self-esteem levels (1948).

At the pinnacle of Maslow's hierarchy of motivational needs is the desire for self-actualization as the employee seeks to reach their full potential driven by intrinsic needs for satisfaction, as the employee's need in health, safety, bonding, and respect are met (Maslow, 1948). Herzberg compliments with the theory of hygiene factors[1] in motivational theory seeking to explain the demotivation, or stepping down Maslow's hierarchy, as lower level needs are raised regarding self-esteem, and social needs (1965; 1948). With this said, human resource managers need to establish policies that support the basic needs, and establish policy that allows for greater motivation in personal and career growth to support the increasing motivational needs of their personnel.

Once effective policies are in place, they must be institutionalized in such a way as to inform and support the project team. Equity in the way in which the information is shared, and opportunity is provided, however, is essential in maintaining what is gained through the establishment of these policies. If in fact the policies exist that provide opportunity for training as an example, removal of such an opportunity, or unreasonably difficult access to the training may become what Herzberg describes as a hygiene issue.

Training and Development

Maslow and Herzberg form a solid basis for further understanding of employee motivational research in the twenty-first century as managers seek to understand motivational theory in relation to their virtual project teams. Herzberg's assertion that the personal rewards provided by the meaningfulness of the employee's work contribution plays heavily in the equation regarding motivational theory and the virtual project team (1965). Contributing to this conversation, Blaskova (2009) exclaims that team member motivation is often formed intrinsically through a worker's internal belief system.

---

1   Herzberg proposed hygiene factors known as the Two-Factor Theory as a way of describing the need to maintain opportunity within the work place such as the pursuit of higher order needs in Maslow's Hierarchy of Needs theory of motivation. At its most basic level, Herzberg's Two-Factor theory may be described to say an opportunity is a motivator as long as it has real potential to be fulfilled, and once fulfilled it then is no longer a motivational factor, and may soon become a demotivator if a new opportunity for fulfillment is not represented in its place.

Workers, Blaskova continues, are willing to adjust their performance based on the presence of motivational factors in the work place, which during times of economic challenges such as those facing managers in the twenty first century are often difficult to provide (2009). As companies look for ways to respond to the extended downturn, and seek ways to address the financial difficulties faced in a small world economy, maintaining a motivational work place is becoming more difficult as teams become increasingly geographically dispersed. As purses are tightened in the face of economic uncertainty, employees are challenged with potential concerns regarding employment security at the lower social and safety needs of Maslow's pyramid (1948).

Powell (2000) expressed this concern declaring that as teams move to the virtual team model, socialization of teams is reduced, hindering the establishment of support and membership that employees once gained through work place relationships. Managers need to find ways in which the rewards are maintained as budgets decline. One way in which I have seen companies make this shift is through creative use of training opportunities.

As managers do, I have had many talks with peers throughout the financial, political, and nuclear power industries. Everyone is struggling with the same challenges. How to stretch the training dollars and maintain an environment that allows for creativity and opportunity in the work force? Many managers are training subject matter experts (SMEs) on a needed topic, and including training and mentorship on how to train their peers on the subject. These newly trained trainers are then offered the opportunity to provide a free lunch for everyone that comes to their training session.

Companies are going back to the days of the total quality management (TQM) craze in order to find new ways to motivate teams through process change as well. Quality Circles are formed to find ways to stretch their metaphorical envelope and build in new motivational opportunities. Employee teams, using employee formed leadership opportunities, are able to use their own creativity to improve how they get work done, develop new processes, and train their peers, allowing for greater opportunity in the higher tiers of Maslow's hierarchy.

Leadership Support of Virtual Teams

Zigurs describes the complexity of leadership in a virtual setting to the degree to which the team is virtual (2003). As more elements of virtuality are added to the team dynamics, the greater the complexity of leadership issues the team may face. Complicating the issues in team complexity for leaders is the issues in team dynamics as members come and go within the team depending upon the work in progress, and changes in project goals and requirements, as well as the necessity for team member autonomy is a virtual setting. Additionally, traditional challenges such as monitoring team progress, providing feedback, and resolving conflicts become greater challenges when working apart from the team.

A natural difference between virtual teams and traditional teams is not simply the physical separation, but the additional challenge of the psychological dispersion. Leaders need new skills in this environment. They become facilitators rather than directive, and Encourages rather than managers (DeRosa, Hantula, Kock & D'Arcy, 2004). These new skills may threaten less experienced managers, and will likely require new training opportunities to help managers cope with the lack of control a virtual environment offers to the management ranks.

Self-identification and categorization as a leader, possibly even more so than as a member of a group, is made more difficult when separation from a group may cause a leader to feel some level of competition for control. Managers that rely upon transactional behaviors as well as those that prefer transformational leadership styles may also face greater challenges in this setting depending upon the degree to which the team may work virtually.

Transactional management styles rely upon the exchange of action and reward, or perhaps management by exception, while transformation leaders may prefer forms of inspiration and extrinsic motivations to move team members to greater levels of performance (Hambley, O'Neill & Kline, 2007). Both styles may be effective, and possibly dependent upon the degree to which teams may operative autonomously. As Spillane (2005) noted, the degree to which autonomy is necessary may increase the need for the distribution of leadership roles to ensure effective team performance. The needs of the team at a point in the project, and possibly the degree to which the diffusion of leadership roles is necessary, may have an impact on which leadership style is effective at any given time.

As Hambley et al. describe the changes in leadership based on situational awareness, teams need different kinds of leaders as the focus of the work shifts (2007). Teams working in more creative roles at some point in the project may need a transformational leader to inspire greater levels of creative thinking and idea development. As the project shifts to a greater level of technical difficulty and engineering functions may be better led by leaders with transactional skills. Short term, fast acting, high performance teams may feel unfairly intruded upon by leaders that inspire to lead teams to greater levels of performance.

Team Collaboration Mediation

Moore (2007) addresses the issue of the psychology of team socialization indicating the need for leadership motivational practices, such as goal setting, personnel development, and activity coordination, is absolutely essential in strengthening virtual, geographically dispersed, teams. Having to work through technologies such as web messaging, text, email, and other forms of communication interventions, Moore suggests that team leaders be required to have the skill sets to build the relationships between team members, create the psychologically safe work environment, and provide encouragement and opportunity for professional growth (2007). In virtual team settings, even more so than in traditional teams, team autonomy may have a greater impact on team collaboration than does the presentation of specific technology.

Peters and Manz (2007) suggest that moving from a traditionally regimented leadership style to that of diffused leadership may provide greater opportunity for collaboration due to the level of freedom this style provides. As teams feel free to experiment and make decisions, they may feel greater empowerment to express their ideas and therefore discuss the opportunity that different options may provide. The technology therefore is a necessary enabling element, but not necessarily a driver of virtual team collaboration.

Peters and Manz suggest that virtual team collaboration, like that of the more traditional face to face project team, may be defined by the degree to which team members may share influence and support, as well as participation in team decisions and outcomes. This includes activities such as shared conflict resolution, creativity and experimentation, and the ability to affect direct communication and participation in such events. Team support, conflict, innovation and creativity, and experimentation are all heavily influenced by the depth of the relationships that team members may build within their virtual environment.

Organizational Culture

Team socialization, according to Moore (2007), is a required motivational tool that must be provided. This is accomplished, according to Moore, most effectively in face to face communications, and at minimum on an intermittent schedule, to address the needs that Maslow (1948) described for social level actualization before a team can attain greater satisfaction and motivation (2007). Managers of effective virtual teams rely heavily on understanding the feelings, work problems, issues, and other situational conditions, as well as individual motivations and team member interdependencies to demonstrate situational awareness and empathy for team members. Additionally, the necessity to listen attentively and in a non-evaluative way is heightened due to the absence of face to face communications.

As companies adjust to allow for decentralization of some decision making, and change reporting structures to make communications more direct, companies are finding that changing the culture of the company can at times require changes in the way the company hierarchy is formed. Many companies, Peters and Manz contend, will go so far as to formally change from a hierarchical management structure to a flat and open company structure (2007). The authors state that as the reporting structure is changed to allow for diffusion in decision making and leadership responsibilities changes, so does the culture of the company.

This works in both large and small organizations. My department contains only three full time employees, and a handful of consultants, and yet even with such a small group we work in a diffused leadership style. Although we all talk multiple times each day, and when I say talk, I include text, email, instant messaging (IM), and phone, as well as intermittent face to face, each person is empowered, and yes encouraged, to make their own decisions. My role is that of facilitator to help make their vision within their own zone of control and responsibility to take shape. I simply provide the needed guidance to help my talented team move in the same direction, and form a consistent and cohesive program.

In an environment with multiple international cultures this can be a bit more challenging, and yet diffused leadership and flat organizational structures can be effective. Once again within my small group we do reflect an international strategy. At least half of my team is offshore, and another third is half way across the United States from the remainder of the team. We exist within three time zones every day, and yet work all problems in a collaborative environment that includes multiple organizations with every discussion.

Having team members that have the experience working with multiple cultures does help. The company needs to be aware that multiple cultures exist in every project, and provide the training and monitoring to ensure the work environment remains sensitive to the needs of these work teams. We make sure to ask about holiday schedules when deliverables are requested, and work to ensure we schedule meeting times that are fair and reasonable across the time zones. Cultural awareness needs to be built into the team's charter and supported by company policies wherever necessary.

Standardized Policy and Procedures

Improving virtual team performance may be accomplished by providing teams with an easy way to improve communications, and clarifying roles and expectations while giving teams a means of implementing a lateral structure in virtual teams (Wong & Burton, 2000). Therefore, as Weems-Landingham points out, the manager of virtual work teams must enlist a multitude of media in employing effective, participative listening (2004). Managers must seek to understand the facts in conflict resolution and coaching events, and draw accurate and complete representations of factual events (Turner, 1999). These simple practices provide a means of ensuring equitable policy application, as well as employing the traditional methods of empathy, teambuilding, and participation to bridge the geographic gap and encourage self-identification with core team membership.

Text messaging, IM, email, and other electronic media are generally considered to be effective and essential communications devices and strategies for effective virtual work teams. Using multiple communication channels are necessary and may be effective, but should never be the singular means of communication and team socialization. Face to face is always necessary and should be worked into the schedule on a regular basis. Within the use of effective communications media must also be the message of equitable application of policy and procedure to reinforce member trust among the team and with the corporation.

Further, standardization of policies and procedures will help team members to become acclimated to a project very quickly. As teams learn what is expected of them on a project, these expectations will easily translate into rapid team formation on the next big project. This further reinforces membership in a virtual setting, and provides a line of sight for team members to a much longer horizon to their work relationships.

Team Leader and Team Member Competency

Trust building may take on a calculative nature as opposed to normative as a means of accommodating political differences (Mizrachi et al., 2007). The effective manager of high performing teams, therefore, needs to find creative ways to fill the needs of the employee in safety and socialization if the team is to be a truly effective virtual workforce. As Herzberg points out, a base element in effecting a satisfying job experience is recognizing those aspects of the employee's work that achieve goals and milestones, as well as the basics of the job requirements in a consistent and effective daily manner (1965).

This may require a new set of management competencies. Competencies are those skills or behaviors that would be agreed to be core to the capability of a group or company to compete in their chosen market. Skills such as the ability to build trust or create a learning organization would be new core competencies needed on the management team of a virtual project organization (Holton, 2001). Other areas of competency that will need to be established are those of team building and group dynamics, conflict resolution, and group communication.

In teams that may be include the virtual elements such as diffusion across time zones and cultures the added competencies of cross-cultural communication, process facilitation, and creating and sustaining remote team work will be needed (Holten, 2001). Hertel, Geister, and Konradt break the discussion of competencies into three general categories of cognitive, task oriented, and teamwork related socio-emotional (2005). Attributes such as conscientiousness and integrity cooperativeness, along with solid communication skills and self-management, become essential in both management and employee level participants in a virtual environment.

Integrating a Virtual Team Management Strategy That Will be the Most Effective for Managing Information Technology Development Projects

An integrated strategy should impact each of the seven factors described by Duarte and Snyder (2006), and therefore each is addressed in the following paragraphs. Management strategy must be organized such that it draws the employee to the organization, fostering commitment, and fulfilling the expectations of the employee engendering trust, openness, and a high level of participation (Kanter, 1968; Zand, 1972). It must be open and participative, allowing for the need of an autonomous element and leadership opportunity, and provide a collaborative environment that encourages synchronous planning and regular closed loop communications utilizing effectively integrated technologies.

Processes should be developed that are based around how work gets done, and supports the remote team members need for participation and collaboration. Remote team members need to feel that their voice is heard, and their needs are accounted for on a daily basis. The perception, as well as the reality, of collaboration and membership in geographically dispersed teams is essential to the ability to build team effectiveness and performance. I say this in this backward fashion for a purpose. Team members working remotely or in a virtual setting need to feel and understand that their voice is heard. Just because the reality may be that their needs have been accounted in the process, does not mean that they may well understand this reality.

Attempting to incorporate the needs of virtual team members, if not managed well, can also be an impediment in enacting effective technology. This may occur in situations where one or more groups are at odds regarding goals, or conflicting social norms. This situation may be aggravated in situations where, as Schwarz and Watson point out, management is in disagreement with other employee groups or IT implementation teams (2005).

While this may sound odd or even contrived, I have often seen technology decisions made to satisfy management desires rather than the needs of the virtual team members. Simple technologies like defect management for IT projects can either support the needs of the virtual team members in information discovery, or serve to aggravate their feelings of membership. I have seen cases where a project team working across work teams at the program level struggle to work in a virtual manner because their request to combine defect management instances of the same tool were ignored. The team members repeatedly requested multiple teams to combine their use of the defect management tool to allow them to better understand the data in a project.

The teams were forced to spend hours attempting to pull information from the tool from multiple instances and combine the data using spreadsheets in order to be able to report on the project progress. To successfully accomplish this, the team had to put one person on data collection and reporting full time.

Requests such as this would seem to be simple, however I saw in one case where the discussion lasted for years.

While the team members continued to struggle in finding and aggregating project information managers continued to argue about their desire for technological freedom to work in a way that was best for their own needs rather than finding a way to accommodate the needs of those working virtually on the team program. This can occur with time tracking, just as easily if multiple instances of the same tool are used in different ways across programs. While these examples may be simple, they do affect perceptions of membership within the team.

HUMAN RESOURCE POLICY

Sarker et al. (2003) and Mizrachi et al. (2007) each discuss at length the need for equitable application of policies and procedures, and the expectation that such policy will consistently provide an environment with little ambiguity regarding organizational goals. Building a participative and open organizational policy requires the active management of knowledge and policy, and successful implementation of a participative policy requires the active participation of all members of the organization up and down the hierarchy (Ardichvili, Page & Wentling, 2003).

Participation, Ardichvili et al. (2003) found, rises from a belief that sharing furthers the greater good of the whole. With this in mind, as policy is seen as equitable and flexible, and as managers seek to bring about fairness and reasonableness regarding home and work life balance, employees identify with a policy of inclusiveness. As employees feel that their voice is heard and their participation has made a difference in corporate policy, they may be more inclined to participate and share in the greater good of the organization, and commit themselves to the success of the whole (Ryan & Kossek, 2008).

Policies that commit to the employee a desire to implement programs that support participation may also feed into the discussion regarding motivational strategies espoused by Maslow and Herzberg. As employees participate in building effective human resource policy that includes an understanding of a virtual work environment, their participation in further problem solving and program development feeds greater opportunity in the higher level motivational factors.

ORGANIZATIONAL CULTURE

As employees and managers gain cultural awareness and a global perspective through the process of virtualization in the organization, the culture of the organization will shift to one of global awareness. An often found side effect of globalization is the added bonus of becoming a learning organization, open to new experiences, and drawing employees and managers closer together in sharing relationships. Openness and acceptance to the differences each employee offers to the organization brings about new levels of sharing, and an environment which allows for greater self-disclosure, forming a culture in which safety, trust, and a team identity may form.

This often happens as managers and employees must learn to lean upon each other for their experiences with other cultures. In my work group alone we have employees that have recently arrived from India, Pakistan, France, Russia, Germany, Ireland, and England, to name only a few of the represented cultures. In addition to these nations, corporate travels have included additional trips to Argentina, Mexico, Romania, Taiwan, Singapore, as well as several Nordic nations as we reach out for greater awareness and understanding of the cultures with whom we work. In order to accomplish this, our organization must work with employees and managers alike to ensure a good understanding of the cultures in order to bring about strong working relationships.

As the corporate culture grows to that of a global family, social awareness and identification with the company and team may grow, strengthening a perception of equitable application of corporate and work group policy (Lipponen et al., 2004). This may then in turn strengthen one's pride in self and team, potentially increasing a sense of safety and acceptance and producing the necessary environment to allow vulnerability for establishment of trust. As Elving and Halgin have described for us, a sense of family will bring to the organization an expectation for an enduring relationship and a heightened sense of belonging and commitment (2005; 2009).

We all realize that not every corporation or company can realistically move employees around the globe as a means of engendering cultural awareness and the growth of familiarity. We can however take advantage of available technology in video conferencing. This technology is often available for a minimal monthly cost, and an inexpensive laptop.

TRAINING AND DEVELOPMENT

Job sharing for key individuals in key positions as a way building a strong understanding of the company, of the many corporate and individual cultures represented in each of these groups, and of the roles and goals of dependent organizations is a solid and proven means of building relationships among groups and teams. As employees and managers shift roles in a job rotation or job sharing process within the organization they will strengthen the effect of information sharing building relationships, and furthering a global perspective of the organization. In addition, the need for fast and effective learning and a newly contrived dependency upon new peers and employees, may bring with it an opportunity for peer training and mentoring, and a greater extension of openness and trust not previously available to virtual work teams. As Abernethy, Piegari, and Reichgelt (2007) note, the experience of mentored, guided training, lends itself well to virtual work teams as the trainer is able to adjust to the needs of the trainee based on feedback, and therefore better assess the degree to which the training fulfills the goals of the organization.

Peer training in a job rotation program also has the effect of providing employees with greater skill and understanding of the full lifecycle of the product, providing for growth, greater commitment, and self-direction and motivation as the employee is able to take part in the whole of the work (Herzberg, 1965). The need for internal initiative and commitment by the individual may bring about greater opportunity for leadership activities in employees as they engage in global work teams. As employees work in an environment that may require greater isolation from the core team, autonomy, situational and organizational awareness, and leadership skills are essential.

STANDARDIZED POLICY AND PROCEDURES

Standardized policies and procedures provide greater ability to bring employees together, and empower them toward decision making and self-direction (Duarte & Snyder, 2006). Policies and procedures need to be documented in a format that provides for the flexibility to tailor actionable instructions for each project and individual team goals to support rapid and consistent team formation. This will help in establishing the capability to position teams within the greater organization for successful goal attainment (Duarte & Snyder, 2006).

As the organization comes together with a set of standard practices, and team members across the organization learn that they can work in similar ways toward dependent goals, building a set of standard work templates will help in building program stability. Standardization of the templates with which teams document project information in application and product development further creates the ability to bring together team members, providing for rapid team normalization, and greater ability to support the high level of activity that virtual teams often must support.

Utilizing a standard set of policy statements and organizational procedures supports the rapid deployment of virtual teams, and allows for the use of a decentralized work teams as needed, however maintaining balance between control of the process and the attempt to develop autonomy and trust is essential to corporate health (Gassman & von Zedtwitz, 2003; Gallivan, 2001). Standardization also supports institutional equity as the need for virtual teams to characterize and rationalize fairness and equity in practices may be reduced as standardization increases. Corporations must maintain the identity of the organization in order to be able to fully support the needs of globally participating teams.

LEADERSHIP SUPPORT OF VIRTUAL TEAMS

Behaviors such as team coordination and planning, clear definition and separation of responsibility, and autonomy of activity, are effective in preventing overlap, and need to be provided by team leadership. Leadership should also find ways to promote and support a consistent and continuous way to maintain active coordination of socialization activities to provide adequate levels of face to face participation (Moore, 2007; Sternberg & Grigorenko, 1993). Managers also need to find ways in which to ensure that team members have the specific technology they need that is tailored to the needs of their project wherever possible, and the capability to effectively implement the technology provided to them for collaboration (Duarte & Snyder, 2006; May & Carter, 2000).

Additionally, technology must be effectively integrated such that different technologies are compatible for information sharing, and integrated with that of the greater corporation to support communications (Duarte & Snyder, 2006; May & Carter, 2000). While this seems like a simple statement, I have seen in large organizations times where decentralized control has allowed too much independence in the area of technology choice. Such decisions as which web based meeting software to use need to be centralized to prevent different groups from making independent decisions that may limit collaboration. At the same time, these decisions need to be integrated with other technology decisions to ensure the tools that are chosen will operate correctly with regionally determined applications.

TEAM LEADER AND TEAM MEMBER COMPETENCY DEVELOPMENT

Moore (2007) suggests that leadership competency regarding establishment of clear, concise, and achievable goals is essential to the success of virtual teams. Active engagement with the team members to provide timely feedback and effective guidance is necessary to identify roadblocks and prevent conflict, as well as supporting team members to come together to resolve discrepancies and maintain forward motion in the face of blocking issues and other unforeseeable challenges. Team members and management alike must be capable of building relationships across cultural and organizational verticals, and geographic boundaries (Duarte & Snyder, 2006). Skills in project management, relationship networking, electronic communication mediation tools, and personal boundary awareness are essential skills for virtual team members to develop and establish as norms within the virtual setting (Duarte & Snyder, 2006).

TEAM COLLABORATION MEDIATION

Andres (2002) suggests that studies consistently reflect the need for electronic mediation in communications as a team building tool in order that decision making may be consistently distributed throughout the team to support team member participation. Teams such as those engaged in collaborative activities requiring high levels of participative mediation to support project processes may require media rich technology such as video conferencing.

In a recent study Researchers noted that trust, either personality based, institutionally based, or cognitively based, are all positively influenced by the use of video conference technology. Teams that were found to be inconsistent in the use of video conference technology, however, were found to have varying degrees of success in improving team member trust that seemed to correlate with the degree of trusting behavior. Younger team members' awareness of team member needs, and their identification with the team may not be as positively correlated to the high frequency of video conference use as are the older team members (Karpiscak, 2007).

Chat rooms, asynchronous blogging and feedback, posting and pulling information from peers, and peer to peer pressure and mentoring of participation in community to share knowledge are active and passive ways in which peer relationships may contribute to knowledge distribution and employee engagement in peer mentoring (Ardichvili, et al., 2003; Hale, 2000; Liu & Batt, 2010). The integration of the electronic work space according to May and Carter must incorporate all of the electronic enablers into one system of communication and information sharing (2000).

Bibliography

Abernethy, K., Piegari, G. & Reichgelt, H. (2007). Teaching project management: An experiential approach. *Journal of Computing Sciences in Colleges.* 22(3), 198–205. Retrieved from http://www.acm.org.

Ahuja, J. (2010). A study of virtuality impact on team performance. *The IUP Journal of Management Research.* IX(5), 27–56. Retrieved from http://proquest.umi.com

Al Qahtani, M. & Daneshgar, F. (2010). An evaluation of the McKinsey's offshoring framework for supply chain relationships. *The Business Review, Cambridge.* 16(1), 178–184. Retrieved from http://www.jaabc.com/.

Altman, I. & Haythorn, W.W. (1965). Interpersonal exchange in isolation. *Sociometry.* 28(4), 411–426. Retrieved from SocINDEX database.

Anantatmula, V. & Thomas, M. (2010). Managing global projects: A structured approach for better performance. *Project Management Journal.* 41(2) 60–72. DOI: 10.1002/pmj.20168.

Andres, H.P. (2002). A comparison of face-to-face and virtual software development teams. *Team Performance Management: An International Journal.* 8(1/2), 39–48. Retrieved from http://www2.hawaii.edu/.

Ardichvili, A., Page, V. & Wentling, T. (2003). Motivation and barriers to participation in virtual knowledge-sharing communities of practice. *Journal of Knowledge Management.* 7(1), 64. Retrieved from http://elearning.ice.ntnu.edu.

Armstrong, D.J. & Cole, P. (2002). *Managing distances and differences in geographically distributed work groups.* [online Google Books]. Retrieved from http://books.google.com

Arney, D. (2011). *Measuring the CQA Processes.* [Unpublished manuscript].

Aron, R. & Singh, J. (2005). Getting offshoring right. *Harvard Business Review.* 83(12), 135–143. Retrieved from http://www.hbr.org.

Baldwin, M.W. (1992). Relational schemas and the processing of social information. *Psychological Bulletin.* 112(3), 461–484. DOI: 10.1037/0033-2909.112.3.461.

Bargh, J.A., McKenna, K.Y. & Fitzsimons, G.M. (2002). Can you see the real me? Activation and expression of the "true self" on the Internet. *Journal of Social Issues*. 58(1), 33–48.

Bergiel, B.J., Bergiel, E.B. & Balsmeier, P.W. (2008). Nature of virtual teams: A summary of their advantages and disadvantages. *Management Research News*. 31(2), 99–110. DOI: 10.1108/01409170810846821.

Bhattacharya, R., Devinney, T.M. & Pillutla, M.M. (1998). A formal model of trust based on outcomes. *Academy of Management Review*. 23(3), 459–472. Retrieved from http:// www.aomonline.org.

Blaskova, M. (2009). Correlations between the increase in motivation and increase in quality. *E+M Ekonomie a Management*. 4, 54. Retrieved from www.em.kbbarko.cz/.

Brenner, J. (2012). Pew Internet: Social Networking (full detail). *Pew Internet & American Life Project*. Retrieved from http://pewinternet.org.

Brooke, C. (1994). Information technology and the quality gap. *Employee Relations*. 16(4) 22. DOI: 10.1108/01425459410066265.

Cocchiara, F.K. & Quick, J.C. (2004). The negative effects of positive stereotypes: ethnicity-related stressors and implications on organizational health. *Journal of Organizational Behavior*. 25(6), 781–785. DOI: 10.1002/job.263.

Cooper, J.D. (1978). Corporate level software management. *IEEE Transactions on Software Engineering*. SE-4(4), 319–326. DOI: 10.1109/TSE.1978.231518.

Cozby, P.C. (1973). Self-disclosure: A literature review. *Psychological Bulletin*. 79(2), 73–91. Retrieved from http://www.apa.org.

Cramm, S.H. (2001). Software insanity. *CIO*. 15(2), 64,75. Retrieved from www.cio.com

de Cremer, D. & Tyler, T.R. (2005). Am I respected or not?: Inclusion and reputation as issues in group membership. *Social Justice Research*. 18(2), 121–153. DOI: 10.1007/s11211-005-7366-3.

Crispin, L. & Gregory, J. (2009). *Agile Testing: A Practical Guide for Testers and Agile Teams*. Upper Saddle River, NJ: Addison-Wesley.

Curry, N. & Fisher, R. (2012). The role of trust in the development of connectivities amongst rural elders in England and Wales. *Journal of Rural Studies*. 1(13). Retrieved from www.elsevier.com.

de Mast, J. (2004). A methodological comparison of three strategies for quality improvement. *The International Journal of Quality & Reliability Management*. 21(2/3), 198–213. Retrieved from www.emeraldinsight.com.

Deming, W.E. (1976). On variances of estimators of a total population under several procedures of sampling. In W.J. Ziegler (eds) *Contributions to Applied Statistics*. Basel and Stutgart: Birkhauser Verlag.

DeRosa, D.M., Hantula, D.A., Kock, N. & D'Arcy, J. (2004). Trust and leadership in virtual teamwork: A media naturalness perspective. *Human Resources Management.* 43(2), 219–232. Retrieved from http://www.wiley.com.

Dossett, D.L. & Hulvershorn, P. (1983). Increasing technical training efficiency: Peer training via computer-assisted instruction. *Journal of Applied Psychology.* 68(4), 552–558. Retrieved from http://www.apa.org.

Duarte, D.L. & Snyder, N.T. (2006). Critical success factors. *Mastering Virtual Teams: Strategies, Tools, And Techniques That Succeed.* [online Google Books]. Retrieved from http://static.managementboek.nl.

Ducheneaut, N., Yee, N., Nickell, E. & Moore, R.J. (2006*). "Alone Together?" Exploring the Social Dynamics of Massively Multiplayer Online Games.* Paper presented at CHI 2006 Proceedings of the Conference on Human Factors and Computing Systems, Montreal, Quebec, Canada. Retrieved from http://www.chi2006.org/cppf.php.

Elving, W.J.L. (2005). The role of communication in organisational change. *Corporate Communications.* 10(2), 129. Retrieved from http://www.corpcommsmagazine.co.uk/.

Fabiansson, C. (2007). Young people's perception of being safe- globally & locally. *Social Indicators Research.* 80(1), 31–49. DOI: 10.1007/s11205-006-9020-3.

Faircloth, B.S. & Hamm, J.V. (2005). Sense of belonging among high school students representing 4 ethnic groups. *Journal of Youth and Adolescence.* 34(4), 293–309. DOI: 10.1007/s10964-005-5752-7.

Fairfax County Virginia Economic Development Authority. (April 2012). *Moving Toward a Non-Traditional Work Force.* Retrieved from www.fairfaxcountyeda.org.

Fiske, S. (1993) Social cognition and social perception. *Annual Review of Psychology.* 44(1), 155. DOI: 10.1146/annurev.ps.44.020193.001103.

Florac, W.A. & Carleton, A.D. (1999). *Measuring the Software Process: Statistical Process Control for Software Process Improvement.* Reading, MA: Addison Wesley Longman, Inc.

Fombrun, C.J. (1996). *Reputation: Realizing Value from the Corporate Image.* Boston, MA: Harvard Business School Press.

Gallivan, M.J. (2001). Striking a balance between trust and control in a virtual organization: A content analysis of open source software case studies. *Information Systems Journal.* 11(4), 277–304. DOI: 10.1046/j.1365-2575.2001.00108.

Gassman, O. & von Zedtwitz, M. (2003). Trends and determinants of managing virtual R&D teams. *R&D Management.* 33(3), 243–262. Retrieved from DOI: 10.1111/1467-9310.00296.

Garza, R.T. & Lipton, J.P. (1978). Culture, personality, and reactions to praise and criticism. *Journal of Personality.* 46(4), 743. Retrieved from http://www.wiley.com.

Gibson, J.W. & Hodgetts, R.M. (1985). Self-disclosure: A neglected skill. *IEEE Transactions on Professional Communications*. PC28(3), 41. Abstract retrieved from http://ieeexplore.ieee.org.

Gilbert, E. & Karahalios, K. (2009). *Predicting Tie Strength With Social Media*. Presented at the proceedings of the 27th international conference on Human factors in Computing Systems, Chicago, IL. Retrieved from http://dl.acm.org.

Gillespie, N.A. & Mann, L. (2004). Transformation leadership and share values: The building blocks of trust. *Journal of Managerial Leadership Psychology*. 19(6), 588–607. DOI: 10.1108/02683940410551507.

Glen, P. (2003). *Leading Geeks; How to Manage and Lead People Who Deliver Technology*. San Francisco, CA: Josey-Bass, A Wiley Inprint.

Godar, S.H. & Ferris, S.P. (2004). Virtual and collaborative teams: Process, technologies and practice. *Trust in Virtual Teams* (Bradley, Vozikis). (pp. 99–114). Hershey, PA: Idea Group Publishing. [online Google Books]. Retrieved from http://books.google.com.

Hakonen, M. & Lipponen, J. (2008). Procedural justice and identification with virtual teams: The moderating role of face-to-face meetings and geographical dispersion. *Social Justice Research*. 21(2), 164–178. DOI: 10.1007/s11211-008-0070-3.

Hale, R. (2000). The science of mentoring at Scottish hydro-electric. *Human Resource Management International Digest*, 8(7), 31. Abstract retrieved from http://www.emeraldinsight.com/.

Halgin, D. (2009). What can managers learn from college basketball? *MITSloan Management Review*. 50(3), 22–24. Retrieved from http://sloanreview.mit.edu/.

Hambley, L. A., O'Neill, T. A. & Kline, T. J. (2007). Virtual team leadership: The effects of leadership style and communication medium on team interaction styles and outcomes. *Organizational Behavior and Human Decision Processes*, 103(1), 1–20.

Haslam, S.A., Postmes, T. & Ellemers, N. (2003). More than a metaphor: Organizational identity makes organizational life possible. *British Journal of Management*. 14, 357–369. Retrieved from http://www.wiley.com.

Hertel, G., Geister, S. & Konradt, U. (2005). Managing virtual teams: A review of current empirical research. *Human Resource Management Review*, 15(1), 69–95.

Herzberg, F. (1965). The new industrial psychology. *Industrial & Labor Relations Review*. 18(3), 364. Retrieved from http://www.ilr.cornell.edu/depts/ilrrev/.

Hinds, P.J. & Bailey, D.E. (2003). Out of sight, out of sync: Understanding conflict in distributed teams. *Organization Science*. 14(6), 615–632. Retrieved from www.jstor.com.

Hinds, P.J. & Mortensen, M. (2005). Understanding conflict in geographically distributed teams: The moderating effects of shared identity, shared context, and spontaneous communication. *Organization Science*. 16(3), 290. DOI: 10.1287/orsc.1050.0122.

Holton, J.A. (2001). Building trust and collaboration in a virtual team, *Team Performance Management*, 7(3), 36–47.

Horenstein, V.D-P. & Downey, J.L. (2003). A cross-cultural investigation of self-disclosure. *North American Journal of Psychology*. 5(3), 373–386. Retrieved from http://www.najp.8m.com.

Jarvenpaa, S.L., Knoll, K. & Leidner, D.E. (1998). Is anybody out there? Antecedents of trust in global virtual teams. *Journal of Management Information Systems*. 14(4), 29. Retrieved from https://digitalcommons.georgetown.edu.

Jarvenpaa, S.L. & Leidner, D.E. (1998). Communication and trust in global virtual teams. *Organization Science*. 10(6), 791–815. Retrieved from http://jcmc.indiana.edu.

Jarvenpaa, S.L., Shaw, T.R. & Staples, D.S. (2004). Toward contexualized theories of trust: The role of trust in global virtual teams. *Information Systems Research*. 15(3), 250. Retrieved from http://www.informs.org/.

Jönsson, C. & Hall, M. (2002, March). *Communication: An Essential Aspect of Diplomacy*. Paper prepared for 43rd annual ISA Convention, New Orleans, LA. Retrieved from http://www.cuts-citee.org/CDS02/pdf/CDS02-Session7-03.pdf.

Kahai, S.K., Sara, T.S. & Kahai, P.S. (2011). Off-shoring and outsourcing. *Journal of Applied Business Research*. 27(1), 113–121. Retrieved from http://www.cluteinstitute.com.

Kanawattanachai, P. & Yoo, Y. (2002). Dynamic nature of trust in virtual teams. *Journal of Strategic Information Systems*. 11(3), 187–213. Retrieved from http://www.elsevier.com.

Kanter, R.M. (1968). Commitment and social organization: A study of commitment mechanisms in utopian communities. *American Sociological Review*. 33(4), 499–517. Retrieved from http://www.asanet.org.

Kaplan, A.M. & Haenlein, M. (2010). Users of the world, unite! The challenges and opportunities of social media. *Business Horizons*. 53(1), 59–68. Retrieved from http://www.journals.elsevier.com/business-horizons/.

Karpiscak, J. (2007). *The Effects of New Technologies on the Performance of Virtual Teams*. (Doctoral dissertation). Capella University, Minneapolis MN. Retrieved from ProQuest Dissertations and Theses database.

Kerr, N.L. & Kaufman-Gilliland, C.M. (1994). Communication, commitment, and cooperation in social dilemmas. *Journal of Personality and Social Psychology*. 66(3), 512–529. DOI: 10.1037/0022-3514.66.3.513.

Kim, A.J. (2000). *Community Building on the Web: Secret Strategies for Successful Online Communities*, [on line]. Abstract retrieved from http://www.kk.org/tools/page56.pdf.

Kirkman, B.L. & Mathieu, J.E. (2005). The dimensions and antecedents of team virtuality. *Journal of Management.* 31(5), 700–718. DOI: 10.1177/0149206305279113.

Kruchten, P. (2004). *The Rational Unified Process: An Introduction.* [online Google Books]. Retrieved from http://books.google.com.

Larman, C. & Vodde, B. (2008). Feature teams. *Scaling Lean & Agile Development; Thinking and Organizational Tools for Large-Scale Scrum.* [online Google Books]. Retrieved from http://www.craiglarman.com/.

Lenhart, A. (2009). Teens and social media: An overview. *PEW/INTERNET & American Life Project.* Retrieved from http://isites.harvard.edu/fs/docs/icb. topic786630.files/Teens%20Social%20Media%20and%20Health%20-%20 NYPH%20Dept%20Pew%20Internet.pdf.

Lewis, J.D. & Weigert, A. (1985). Trust as a social reality. *Social Forces.* 63(4), 967–985. Retrieved from http://socialforces.unc.edu/.

Lipnack, J. & Stamps, J. (1997). *Virtual Teams: Reaching Across Space, Time and Organizations with Technology.* New York: John Wiley and Sons. [online Google Books]. Retrieved from http://books.google.com.

Lipponen, J., Olkkonen, M.-E. & Myyry, L. (2004). Personal value orientation as a moderator in the relationships between perceived organizational justice and its hypothesized consequences. *Social Justice Research.* 17(3), 275–292. Retrieved from http://www.springer.com.

Liu, X. & Batt, R. (2010). How supervisors influence performance: A multilevel study of coaching and group management in technology-mediated services. *Personnel Psychology.* 63(2), 265. http://www.wiley.com.

Martens, A., Johns, M., Greenberg, J. & Schimel, J. (2006). Combating stereotype threat: The effect of self-affirmation on women's intellectual performance. *Journal of Experimental Social Psychology.* 42(2), 236–243. DOI: 10.1016/j.jesp.2005.04.010.

Maslow, A.H. (1948). Some theoretical consequences of basic need-gratification. *Journal of Personality,* June, 402–416. DOI: 10.1111/j.1467-6494.1948.tb02296.x.

May, A. & Carter, C. (2000). A case study of virtual team working in the European automotive industry. *International Journal of Industrial Ergonomics.* 27(3), 171–186. Retrieved from http://www.sciencedirect.com.

Maznevski, M.L. & Chudoba, K.M. (2000). Bridging space over time: Global virtual team dynamics and effectiveness. *Organization Science.* 11(5), 473–492. Abstract retrieved from http://orgsci.journal.informs.org/.

McManus, J. (2004). A stakeholder perspective in software project management. *Management Services.* 48(5), 8–12. Retrieved from www.ims-productivity.com.

McManus, J. & Wood-Harper. T. (2007). Software engineering: A quality management perspective. *The TQM Magazine.* 19(4), 315–327. Retrieved from http://www.emeraldinsight.com.

Merriam-Webster. (October, 2010). [online]. Retrieved from http://www. merriam-webster.com/dictionary/quality%20assurance.

Millward, L.J., Haslam, S.A. & Postmes, T. (2007). Putting employees in their place: The impact of hot desking on organizational and team identification. *Organization Science*. 18(4), 547–559. Retrieved from http://orgsci.journal. informs.org/.

Mizrachi, N., Drori, I. & Anspach, R.R. (2007). Repertoires of trust: The practice of trust in multinational organization amid political conflict. *American Sociological Review*. 72(1), 143–165. Retrieved from http://www.asanet.org.

Montoya-Weiss, M.M., Massey, A.P. & Song, M. (2001). Getting it together: Temporal coordination and conflict management in global virtual teams. *Academy of Management Journal*. 44(6), 1251–1262. Retrieved from http:// aomonline.org/.

Moore, T. G. Jr. (2007). *Virtual Team Member Motivation in New Product Development: An Investigation into the Influence of Leadership Behaviors*. (Doctoral dissertation). Capella University, Minneapolis, MN. Retrieved from ProQuest Dissertations and Theses database.

Morris, M.G. & Venkatesh, V. (2000). Age differences in technology adoption decisions: Implications for a changing workforce. *Personnel Psychology*. 53(2), 375–403. Retrieved from http://www.ouderenenarbeid.uhasselt.be/ Documenten/artikel%20Morris%20Venkatesh.pdf

Nardi, B.A., Whittaker, S. & Bradner, E. (2000). *Interaction and Outeraction: Instant Messaging in Action*. Paper presented at the ACM 2000 Conference on Computer Supported Cooperative Work, Philadelphia, PA. DOI: 10.1145/358916.358975.

Panteli, N. & Duncan, E. (2004). Trust and temporary virtual teams: Alternative explanations and dramaturgical relationships. *Information Technology & People*. 17(4), 423–441. Retrieved from http://www.emeraldinsight.com.

Park, S., Henkin, A.B. & Egley, R. (2005). Teach team commitment, teamwork and trust: Exploring associations. *Journal of Educational Administration*. 43(4/5), 462. Retrieved from http://www.emeraldinsight.com.

Pavlou, P.A. (2002). Institution-based trust in interorganiational exchange relationships: The role of online B2B marketplaces on trust formation. *Journal of Strategic Information Systems*. 11(3–4), 215–243. Retrieved from www.elsevier.com.

Perry, W.E. & Rice, R.W. (1997). *Surviving the Top Ten Challenges of Software Testing: A People-Oriented Approach*. New York, NY: Dorset House Publishing.

Peskin, M.P. & Hart, J. (1996). Measuring the quality of computer systems development. *Benchmarking for Quality Management & Technology*. 3(2), 68. Retrieved from http://www.emeraldinsight.com.

Peters, L.M. & Manz, C.C. (2007). Identifying antecedents of virtual team collaboration. *Team Performance Management*, 13(3/4), 117–129.

Polzer, J.T., Milton, L.P. & Swann, W.B.Jr. (2002). Capitalizing on diversity: Interpersonal congruence in small work groups. *Administrative Science Quarterly*. 47(2), 296–324. Retrieved from http://www.johnson.cornell.edu.

Postmes, T., Spears, R. & Lea, M. (2002). Intergroup differentiation in computer-mediated communication: Effects of depersonalization. *Group Dynamics: Theory, Research, and Practice*. 6(1), 3–16. DOI: 10.1037/1089-2699.6.1.3.

Powell, A.L. (2000). *Antecedents and Outcomes of Team Commitment in a Global, Virtual Environment*. (Doctoral dissertation). Indiana University, Indiana.

Powell, A., Piccoli, G. & Ives, B. (2004). Virtual Teams: A review of current literature and directions for future research. *The DATA BASE for Advances in Information Systems*. 35(1), 6–36. Retrieved from http://www.sigmis.org.

Powell, W.W. (1990). Neither market nor hierarchy: Network forms of organization. *Research in Organizational Behavior*, 12, 295–336. Retrieved from http://www.stanford.edu.

Purdy, J.M., Nye, P. & Balakrishnan, P.V. (2002). The impact of communication media on negotiation outcomes. *The International Journal of Conflict Management*. 11(2), 162–187. Retrieved from www.emeraldinsight.com/ijcma.html.

Reilly, J.P. (1995). Does RAD live up to the hype. *IEEE Software*. 12(5), 24. DOI: 10.1109/52.406752.

Roberts, J. (2005). Transparency and self-disclosure. *Family Process*. 44(1), 45–63. DOI: 10.1111/j.1545-5300.2005.00041.x.

Rotter, J.B. (1971). Generalized expectancies for interpersonal trust. *American Psychologist*. 26(5), 443–452. Retrieved from http:// http://www.apa.org/.

Ryan, A.M. & Kossek, E.E. (2008). Work-life policy implementation, breaking down or creating barriers to inclusiveness. *Human Resource Management*. 47(2), 296–310. Retrieved from http://ellenkossek.lir.msu.edu/documents/07HRM47_2ryan.pdf.

Sarker, S., Valacich, J.S. & Sarker, S. (2003). Virtual team trust: Instrument development and validation in an IS educational environment. *Information Resources Management Journal*. 16(2), 35. Retrieved from http://igi-global.com.

Schlenker, B.R. (1975). Liking for a group following an initiation: Impression management or dissonance reduction? *Sociometry*. 38(1), 99–118. Retrieved from http://www.asanet.org.

Schwarz, G.M. & Watson, B.M. (2005). The influence of perceptions of social identity on information technology-enabled change. *Group & Organization Management*. 30(3), 289. DOI: 10.1177/1059601104267622.

Shaw, J.B. & Barrett-Power, E. (1998). The effects of diversity on small work group processes and performance. *Human Relations.* 51(10), 1307–1325. Retrieved from www.sagepublications.com.

Spillane, J.P. (2005). Distributed leadership. *The Educational Forum.* 69(2), 143–150. DOI: 10.1080/00131720508984678.

Sternberg, R.J. & Grigorenko, E. (1993). Shared mental models in expert team decision making. In N.J. Castellan, Jr. (Ed.) *Individual and Group Decision Making.* (Cannon-Bowers, Salas & Converse). (pp. 221–230). Hillsdale, NJ: Lawrence Erlbaum Associates, Inc. Retrieved from http://books.google.com.

Talby, D., Keren, A., Hazzan, O. & Dubinsky, Y. (2006). Agile software testing in a large-scale project. *IEEE Software.* 23(4), 30. DOI: 10.1109/MS.2006.93.

Taylor, J.K. (1987). *Quality Assurance of Chemical Measurements.* Boca Raton, FL: CRC Press LLC. Retrieved from http://books.google.com/.

Turner, J.C. (1999). Some current issues in research on social identity and self-categorization theories. In N. Ellemers, R. Spears & B. Doosje (Eds), *Social Identity* (pp. 6–34). Malden, MA: Blackwell Publishers, Inc. [online Google books]. Retrieved from http://books.google.com.

United States Department of Labor, Bureau of Labor Statistics. Precis, *Monthly labor review online, February 2010.* (February, 2010). Retrieved from http://www.bls.gov.

United States Department of Labor, Bureau of Labor Statistics, Occupational outlook handbook, 2010–11 Edition. (February 14, 2011). *Computer Systems Analysts.* Retrieved from http://www.bls.gov.

United States Department of Labor, Bureau of Labor Statistics, TED: The Editors Desk. *Employed Foreign-Born and Native-Born Persons by Occupation, 2010.* (June 01, 2011). Retrieved from http://www.bls.gov.

United States Department of Labor, Bureau of Labor Statistics, TED: The Editors Desk. *Labor Force Projections to 2020.* (February 16, 2012). Retrieved from http://www.bls.gov.

Wan, CS. & Chiou, WB. (2006). Psychological motives and online games addiction: A test of flow theory and humanistic needs theory for Taiwanese adolescents. *CyberPsychology & Behavior.* 9(3), 317–324. Retrieved from www.nsc.gov.tw.

Watkins-Allen, M., Armstrong, D.J., Riemenschneider, C.K. & Reid, M.F. (2006). Making sense of the barriers women face in the information technology work force: Standpoint theory, self-disclosure, and causal maps. *Sex Roles.*54(11), 831–844. DOI: 10.1007/s11199-006-9049-4.

Weems-Landingham, V. *The role of Project Manager and Team Member Knowledge, Skills and Abilities (KSAs) in Distinguishing Virtual Project Team Performance Outcomes,* (Doctoral dissertation), 2004, Capella University, Minneapolis, MN.

Weyuker, E.J., Ostrand, T.J., Brophy, J. & Prasad, R. (2000). Clearing a career path for software testers. *IEEE Software.* 17(2), 76. DOI: 10.1109/52.841696.

Whetten, D.A. & Cameron, K.S. (1995). *Developing Management Skills* (3rd ed.). New York, NY: HarperCollins College Publishers.

Whitener, E.M., Brodt, S.E., Korsgaard, M.A. & Werner, J.M. (1998). Managers as initiators of trust: An exchange relationship framework for understanding managerial trustworthy behavior. *The Academy of Management Review.* 23(3), 513. Retrieved from http://www.aomonline.org.

Wilkens, R. & London, M. (2006). Relationships between climate, process, and performance in continuous quality improvement groups. *Journal of Vocational Behavior.* 69(3), 510–523. Retrieved from http://www.elsevier.com.

Wise, T.P. (2011a). Project team socialization: Are text messaging and IM damaging team performance? *The Journal for Quality and Participation.* April 2011. 34(1). [online].

Wise, T.P. (2011b). Software quality assurance – a simple strategy for implementation. *Quality Progress.* 44(6), 30–35. Retrieved from www.asq.org.

Wise, T.P. (2012). *The Effect of Geographical Separation, Mediated Communications, and Culture on Tester Team Member Trust of Other Information Technology Virtual Project Team Members.* (Doctoral dissertation in preparation). Capella University, Minneapolis, MN.

Wong, S-S. & Burton, R.M. (2000). Virtual teams: What are their characteristics, and impact on team performance? *Computational & Mathematical Organization Theory.* 6(4), 339–360. Retrieved from http://springer.com.

Wright, T., Boria, E. & Breidenbach, P. (2002). Creative player actions in FPS online video games: Playing counter strike. *The International Journal of Computer Game Research.* 2(2). Retrieved from www.gamestudies.org.

Yee, N. (2006). The psychology of massively multi-user online role-playing games: Motivations, emotional investment, relationships and problematic usage. *Avatars at Work and Play.* 187–207. DOI: 10.1007/1-4020-3898-4_9.

Zand, D.E. (1972). Trust and managerial problem solving. *Administrative Science Quarterly.* 17(2), 229–239. Retrieved from http://www.johnson.cornell.edu.

Zigurs, I. (2003). Leadership in Virtual Teams:-Oxymoron or Opportunity?. *Organizational Dynamics,* 31(4), 339–351.

Further Reading

Ainsworth, M.S. & Bowlby, J. (1991). An ethological approach to personality development. *American Psychologist.* 46(4), 333–341. Retrieved from http://http://www.apa.org/

Barnes, P.C. (1995). Managing Change. *British Medical Journal.* 310(6979), 590. Retrieved from http://www.bmj.com.

Benyon-Davies, P., Carne, C., Mackay, H. & Tudhope, D. (1999). Rapid application development (RAD): An empirical review. *European Journal of Information Systems.* 8(3), 211–223. Retrieved from http://www.palgrave-journals.com.

Carroll, J. (1995). The application of total quality management to software development. *Information Technology & People.* 8(4), 35. Retrieved from http://www.itandpeople.org/.

Christel, M.G. & Kang, K.C. (1992). *Issues in Requirements Elicitation* (CMU/SEI-92-TR-012, ESC-TR-92-012). Carnegie Mellon University, Software Engineering Institute. Retrieved from http://www.sei.cmu.edu.

Citrelli, J.W. & Neumann, K.F. (1978). An interpersonal analysis of self-disclosure and feedback. *Social Behavior & Personality: an International Journal.* 6(2), 173. Retrieved from http://www.sbp-journal.com.

Côte, M., Suryn, W. & Georgiadou, E. (2007). In search for a widely applicable and accepted software quality model for software quality engineering. *Software and Information Technology Engineering Department, École de technologie supérieure,* Montreal, QC, Canada. DOI: 10.1007/s11219-007-9029-0.

Dac-Buu, C. (2006). *An empirical investigation of critical success factors in agile software development projects.* (Doctoral dissertation in preparation). Capella University, Minneapolis, MN.

Davis, C.G. & Vick, C.R. (1977). The software development system. *IEEE Transactions on Software Engineering, SE-3(1),* 69–84. DOI: 10.1109/TSE.1977.233839.

Deutsch, M. (1960). Critique and notes: Trust, trustworthiness, and the F Scale. *The Journal of Abnormal and Social Psychology.* 61(1), 138–140. DOI: 10.1037/h0046501.

Ellison, N.B., Steinfield, C. & Lampe, C. (2007). The benefits of Facebook "friends:" Social capital and college students' use of online social network sites. *Journal of Computer-Mediated Communication.* 12(4), 1143–1168. DOI: 10.1111/j.1083-6101.2007.00367.x.

Eom, S.B. & Lee, C.K. (1999, Spring). Virtual teams: An information age opportunity for mobilizing hidden manpower. *SAM Advanced Management Journal.* 64(2). Retrieved from http://www.samnational.org.

Folger, R. & Bies, R.J. (1989). Managerial responsibilities and procedural justice. *Employee Responsibilities and Rights Journal.* 2(2). Retrieved from http://www.springer.com.

Guzzo, R.A. & Dickson, M.W. (1996). Teams in organizations: Recent research on performance and effectiveness. *Annual Review of Psychology*. 47(1), 307. Retrieved from http://www.annualreviews.org/.

Hamsher, J.H., Geller, J.D. & Rotter, J.B. (1968). Interpersonal trust, internal-external control, and the Warren Commission report. *Journal of Personality and Social Psychology*. 9(3), 210–215. DOI: 10.1037/h0025900.

Horch, J.W. (1996). *Practical Guide to Software Quality Management*. Norwood, MA: Artech House, Inc.

Iacono, C.S. & Weisband, S. (1997). *Developing Trust in Virtual Teams*. Presented at the proceedings of the 30th annual Hawaii international conference on systems sciences. Retrieved from http://www.hicss.hawaii.edu/.

Johns, M., Schmader, T. & Martens, A. (2005). Knowing is half the battle: Teaching stereotype threat as a means of improving women's math performance. *Psychological Science*. 16(3), 175–179. DOI: 10.1111/j.0956-7976.2005.00799.x.

Joinson, A.N. (2001). Self-disclosure in computer-mediated communication: The role of self-awareness and visual anonymity. *European Journal of Social Psychology*. 31, 177–192. DOI: 10.1002/ejsp.36.

Jussim, L. (2002). Self-fulfilling prophecies. *International Encyclopedia of the Social & Behavioral Sciences*. 13830–13833. DOI: 10.1016/BO-08-043076-7/01731-9.

Kastro, Y. & Bener, A. B. (2008). A defect prediction method for software versioning. *Software Quality Journal*. 16(4), 543–562 DOI: 10.1007/s11219-008-9053-8.

Katz, D. & Kahn, R.L. (1978). *The Social Psychology of Organizations*. New York, NY: Wiley.

Kirkman, B., Benson, R., Gibson, C.B., Tesluk, P.E. & McPherson, S.O. (2002). Five challenges to virtual team success: Lessons from Sabre, Inc. *Academy of Management Executive*. 16(3), 67–79. Retrieved from http://home.sandiego. edu/~pavett/docs/gsba501/Impact_Team_Empower.pdf.

Laurenceau, J-P., Barrett, L.F. & Pietromonaco, P.R. (1998). Intimacy as an interpersonal process: The importance of self-disclosure, partner disclosure, and perceived partner responsiveness in interpersonal exchanges. *Journal of Personality and Social Psychology*. 74(5), 1238–1251. DOI: 10.1037/0022-3514.74.5.1238.

Marks, M.A., Mathieu, J.E. & Zaccaro, S.J. (2001). A temporally based framework and taxonomy of team processes. *The Academy of Management Review*. 26(3), 356–376. Retrieved from http://www.aomonline.org.

Meyer, J.P. & Allen, N.J. (1997). *Commitment in the Workplace: Theory, Research and Application*. Thousand Oaks, CA: Sage Publishers.

Morgan, R.M. & Hunt, S.D. (1994). The commitment-trust theory of relationship marketing. *Journal of Marketing*. 58(3), 20. Retrieved from http://www. marketingpower.com.

Mugridge, R. (2008). Managing agile project requirements with storytest-driven development. *IEEE Software*. 25(1), 68–75. Retrieved from.www.ieeexplore. ieee.org.

NetMBA, (n.d.). *Herzberg's Motivation–Hygiene Theory (Two Factor Theory)*. Retrieved September, 2012, from www.netmba.com/mgmt/ob/motivation/ herzberg/.

Posey, C., Lowry, P.B., Roberts, T.L. & Ellis, T.S. (2010). Proposing the online community self-disclosure model: The case of working professionals in France and the UK who use online communities. *European Journal of Information Systems*. 19(2), 181. Abstract retrieved from http://proquest.umi.com.

Postmes, T., Tanis, M. & de Wit, B. (2001). Communication and commitment in organizations: A social identity approach. *Group Processes and Intergroup Relations*. 4(3), 207–226. DOI: 10.1177/1368430201004003004.

Sarker, S. & Sarker, S. (2009). Exploring agility in distributed information systems development teams: An interpretive study in an offshoring context. *Information Systems Research*. 20(3), 440–461. Abstract retrieved from http:// informs.org/.

Schaefer, H. (1999). Surviving software testing under time and budget pressure. *Telektronikk*. 95(1), 66–72. Retrieved from http://www.telektronikk.com/.

Self, P.C. & Gebhart, K.A. (1980). A quality assurance process in health sciences libraries. *Bulletin of Medical Library Association*. 68(3), 288–292. Retrieved from http://www.ncbi.nlm.nih.gov/pmc/articles/PMC226510/pdf/mlab00079-0050.pdf.

Shechtman, Z., Hiradin, A. & Zina, S. (2003). The impact of culture on group behavior: A comparison of three ethnic groups. *Journal of Counseling and Development*. 81(2), 208. DOI: 10.1002/j.1556-6678.2003.tb00244.x.

Stough, S., Eom, S. & Buckenmyer, J. (2000). Virtual teaming: A strategy for moving your organization into the new millennium. *Industrial Management & Data Systems*. 100(8), 370–378. Retrieved from https://cstl-hcb.semo.edu/ eom/research/virtualteam200293.pdf.

Tajfel, H., Billig, M.G. & Bundy, R.P. (1971). Social categorization and intergroup behavior. *European Journal of Social Psychology*. 1(2), 149–178. Abstract retrieved from http://www.wiley.com.

Townsend, A.M., DeMarie, S.M. & Hendrickson, A.R. (1998). Virtual teams: Technology and the workplace of the future. *Academy of Management Executive*. 12(3), 17–29. Abstract retrieved from http://www.aomonline.org.

United States Department of Labor, Bureau of Labor Statistics, Occupational outlook handbook, 2010–11 Edition. (February 14, 2011). *Computer Software Engineers and Computer Programmers*. Retrieved from http://www.bls.gov.

Vogt, W.P. (2005). *Dictionary of Statistics & Methodology: A Non-Technical Guide for the Social Sciences.* Thousand Oaks, CA: Sage Publications, Inc.

Vogt, W.P. (ed.). (2007). *Quantitative Research Methods for Professionals.* Boston, MA: Pearson Education, Inc.

Vuckovic, A. (2009). Inter-cultural communication: A foundation of communicative action. *Multicultural Education & Technology Journal.* 2(1), 47. DOI: 10.1108/17504970810867151.

Wayne, S.J. & Green, S.A. (1993). The effects of leader-member exchange on employee citizenship and impression management behavior. *Human Relations.* 46(12), 1431. Retrieved from http://www.sagepub.com.

Weil, N. (2007). From here to agility; Study after study indicates that agile methodologies produce better results in software development and project management. So why have so few CIO's adopted them? Read to find out how you can change your organization's culture of development. *CIO.* 20(16). Retrieved from www.cio.com.

Wiesenfeld, B.M., Raghuram, S. & Garud, R. (1998). Communication patterns as determinants of organizational identification in a virtual organization. *Journal of Computer Mediated Communication.* 3(4). DOI: 10.1111/j.1083-6101.1998.tb00081.x.

Zaccaro, S.J., Rittman, A.L. & Marks, M.A. (2001). Team leadership. *The Leadership Quarterly.* 12(4), 451–483. Retrieved from http://www.elsevier.com.

Index

Printed and bound by CPI Group (UK) Ltd, Croydon, CR0 4YY

18/10/2024

01776204-0001